READINGS IN ECOLOGY AND ENVIRONMENTAL SCIENCE

READINGS IN ECOLOGY AND ENVIRONMENTAL SCIENCE

Madhab Chandra Dash

GenNext Publication

First Published, 2018

Published by

GenNext Publication
5, Ansari Road
Daryaganj, New Delhi-110002
Phone: 9811692060
E-mail: gennextpublication@hotmail.com

Printed at: G. Print Process, Delhi.

Cataloging in Publication Data—DK
Courtesy: D.K. Agencies (P) Ltd. <docinfo@dkagencies.com>

Dash, Madhab C., author.
Readings in ecology and environmental science / Madhab Chandra Dash.
pages cm

ISBN: 978-93-5324-119-3

1. Biodiversity—India—Odisha. 2. Global warming—India—Odisha. 3. Climatic changes—India—Odisha. 4. Human ecology—India—Odisha. I. Title.

LCC QH77.I5D37 2018 | DDC 333.95160954133 23

Preface

The compilation 'Readings in Ecology and Environmental Science' contains some selected papers, which I have written or delivered lectures after my superannuation from Sambalpur University as Professor of Life Science in 1999 and completion of my term as Vice Chancellor from 2001-2004. Some of the papers were published earlier in the International Journal 'The Ecoscan'. I wish to thank the editor, Prof.Manoranjan Sinha, Vice Chancellor, of Sido Kanhu Murmu University, Dumka, Jharkhand, for granting permission to include the papers, published earlier in The Ecoscan. Prof.Niranjan Behera of Sambalpur University, Prof K.G.Saxena of JNU, New Delhi and Dr. Ch.Satyabrata Nanda were, my coauthors in three papers and I thank them profusely for their permission to include the papers in this book.

This compilation has been made to provide reading materials to students, research scholars and public at large, who may be interested to know some aspects of contemporary thinking on ecological and environmental sciences.

Madhab Chandra Dash

2nd October, 2017

Contents

Part IV : Global Warming and Climate Change

Part VI : What is Scientific Temper?

Part I : Environmental Education

1

What is Environmental Education?

Introduction

The concept of Environmental education originated in ancient human civilization and people's awareness was based on moral and ethical values, and narration through fables for children. The roots of environmental education can be traced back to Vedic Times' Rakhaye Prakrutim Pantu Loka'.In eastern culture; importance of nature study (first phase) started in Guru Ashrams but did not take a concrete shape to develop a curriculum for the schools and colleges to address the problems. However, nature was considered as the expression of the Almighty. People worshiped Sun, Moon, Planets, Trees, Rivers, and Soil, some animal species etc.

Manusmruti comprises 2684 verses in 12 chapters. The treatise focused on environmental protection and gave a meaningful structure and is deductive. These ethical laws had significant effect in post vedic period to protect trees, forests, animals, and the five gross elements such as the Ether (Aakaassa,), Vaayu (Air), Tejas (Fire) Aaapa (Water) and Kshiti (earth/Soil) (Buhler, G.1886). These laws have ethical and moral values but were not inductive.

In the second phase, Vernadsky of Russia (www.rdasia.com) developed the concept of Biosphere, the zone that supports life on earth and recognised the geological force and described the atmosphere as extension of Life in 1922. James Lovelock (1919-) of England proposed the Gaia hypothesis suggesting that the earth is a single living organism. The biodiversity interact with the earth, each having its small effect on the system as the individual cells cause some effect on human body.Ultimateky the planet is self regulating and maintains its own surface condition. (www.rdasia.com)

The third phase dealt with systems concept and scientists tried to incorporate management aspects of natural resources for human welfare. This led to develop Environmentalism. During the last 200 years, the concepts of nature study (Jean-Jacques Rousseau) and systems approach has developed by the western scholars into two broad aspects such as (i) Rigorous Science, technology and Management aspects; and (ii) sensitization of people about the environmental degradation, their responsibility & role to improve the situation and their rights as per the provisions of the national constitutions and by the national and international environmental laws enacted by the legislative bodies and UNO. The systems approach of study emphasizes quantification of structural and functional aspects of natural systems and also man-made systems and predictability of the change of the systems under different conditions.

The 4th phase of the development of the concept of environmental education gives importance to Ecosystem Analysis (structural change, functional change in normal conditions, under stress factors, ecosystem stability, restoration and sustainability) and modeling and impact of anthropogenic activity and climate change on natural and man-made ecosystems.

At present international and national conventions are playing important role and have influenced educators to frame syllabus and courses befitting to the requirement of the present day. This paper deals with history, definition, progress of environmental education and the need of the study of ecosystem analysis.

Brief History

The roots of environmental education can be traced back to Vedic Times' Rakhaye Prakrutim Pantu Loka'.In eastern culture; importance of nature study started in Guru Ashrams but did not take a concrete shape to develop a curriculum for school. However, nature was considered as the expression of the Almighty. People worshiped Sun, Moon, Planets, Trees, Rivers, and Soil etc. Manusmruti comprises 2684 verses in 12 chapters. The treatise focused on environmental protection and gave a meaningful structure and is deductive. These ethical laws had significant effect in post vedic period to protect trees, forests, animals, and the five gross elements such as the Ether (Aakaassa,), Vaayu (Air), Tejas (Fire) Aaapa (Water) and Kshiti (earth/Soil) (Buhler, G.1886, Padhy, S et al, 2006). These laws have ethical and moral values but were not inductive.

In recent times as early as the 18th century, Jean-Jacques Rousseau stressed importance of an education that focuses on Environment. Several decades later, Louis Agassiz, a Swiss-born naturalist supported Rousseau's philosophy. He encouraged students to 'Study nature'. Both the scholars laid the foundation for environmental education program, known as Nature Study, which happened in the late 19th century and early 20th century.

In 1876, the German Biologist, Ernest Haeckel coined a word 'Ecology', -Eco means House/Home and logus means discourse. Environmental Education has a strong root in understanding ecological principles.

Initially the nature study movement used fables and moral lessons to help school students to develop an appreciation of nature as man is part of the nature (originally Eastern Concept).. Anna Botsford Comstock, the head of the Department of Nature Study at Cornell University, USA, was a prominent figure in the nature study movement and wrote a Handbook for Nature Study in 1911, to educate school children on cultural values. Comstock and other leaders of Nature study movement, such as Liberty Hyde Bailey, involved community leaders, teachers, and scientists and changed the science curriculum for school children across the United States. Vernadsky of Russia (www.rdasia.com) developed the concept of Biosphere, the zone that supports life on earth and recognised the geological force and described the atmosphere as extension of Life. James Lovelock of England proposed the Gaia hypothesis suggesting that the earth is a single living organism. The biodiversity interact with the earth, each having its small effect on the system as the individual cells cause some effect on human body.Ultimateky the planet is self regulating and maintains its own surface condition. (www.rdasia.com)

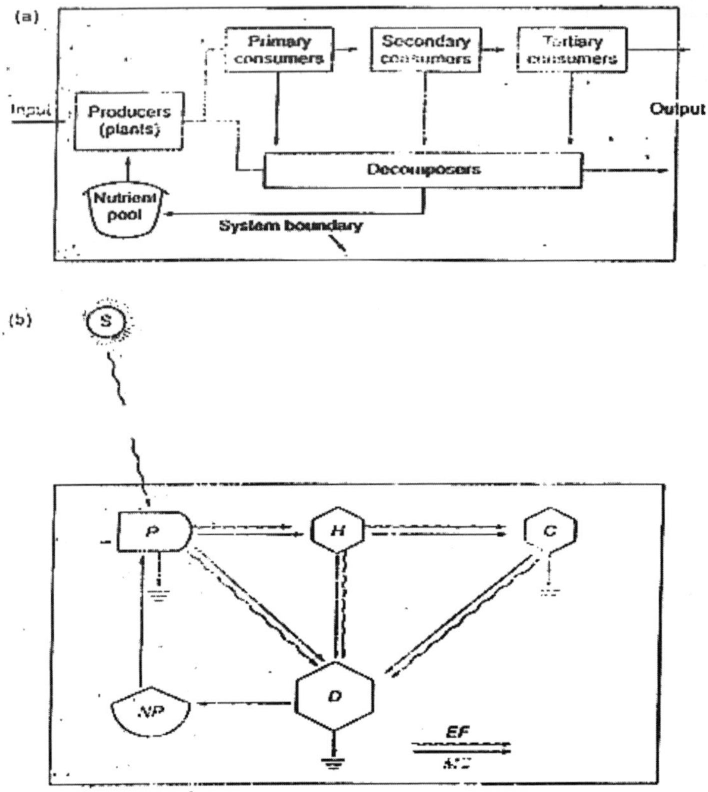

(a) Ecosystem visualised as a single block design (abiotic and biotic components and their interactions are visualised). (b) an ecological system showing system boundary, component parts, energy flow and matter cycling (S = sun, P = producers, H = herbivores, C = carnivores, D = decomposers, NP = nutrient pool, EF = energy flow, MC = matter cycling)

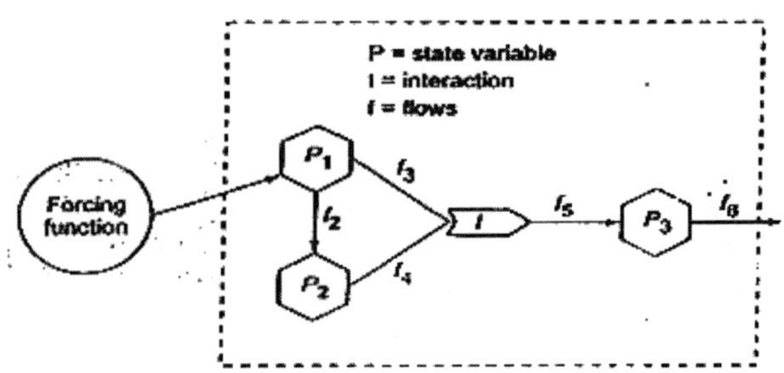

Modelling ecosystems

Figure 1. Ecosystem Concept.

A new type of environmental education, named Conservation Education, emerged during 1920s and 1930s. Conservation Education dealt with natural world and focused on rigorous scientific training involving scientific management and planning tool that helped to solve social, economic, and environmental problems during that time.

In 1935, a British biologist- A.G.Tansley extended the system concept (Ecosystem) developed by William Harvey, -a human physiologist in 16th century and few physical scientists to Ecology, the mother of Environmental Science. (Dash and Dash, 2009).

Figure 1 shows graphical Black Box model, the model showing ecological notations and the model for modeling purpose.

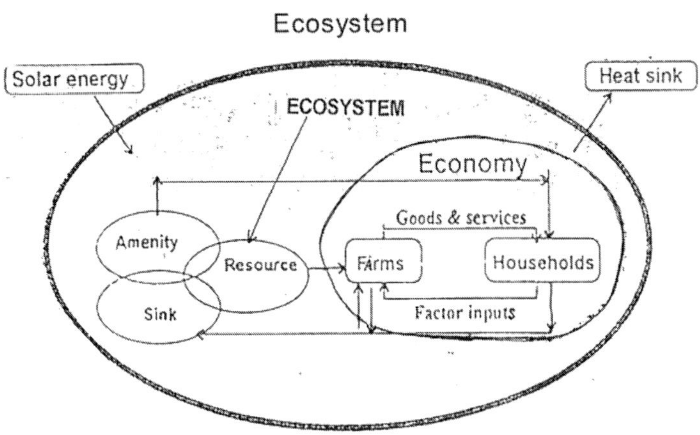

Figure 2. Shows a comparison of natural and man-made economic system.

He thought the living cell is a tiny system and its function is quantified by the resource inputs, biosynthesis, metabolism & the output (growth) etc. The quantification helps to understand the structure of functional dynamics with predictable characteristics. In this way, all type of living cells under different conditions can be quantified and compared. He extended this idea to a mega-system such as an ecological system or Ecosystem. This concept led to develop a new discipline Ecosystem Analysis.

Ecosystems have been classified on the basis of energy flow (Figure 3) showing sun is the source of energy for natural ecosystems and fossil fuel and other form of energy have become the source of energy for man-made systems. Man-made systems are usually monospecies, higher rate of productive, fragile, pollution generating, etc.

Graphical model of systems

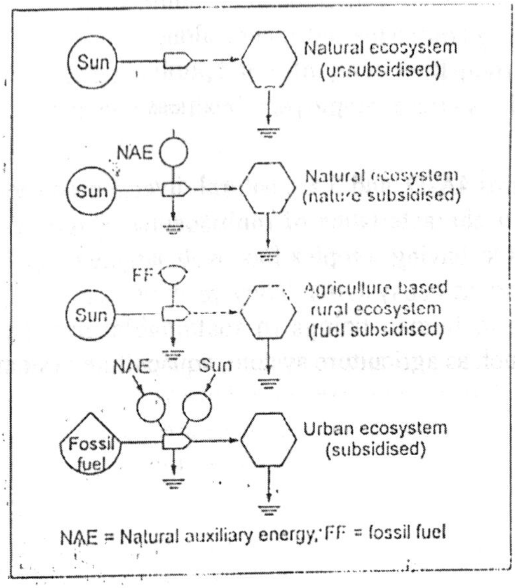

Ecosystems are classified on the basis of energy flow (based on Odum, 1975)

Figure 3. Energy flow basis of ecosystem analysis.

Modern Environmental Education Movement

The modern environmental education movement, gained significant momentum in the late 1960s and early 1970s, as more people began to fear the fallout from radiation, significant amounts of air and water pollution and solid and liquid waste generation, chemical pesticides mentioned in Rachel Carson's 'Silent Spring', the minimata disease due to mercury pollution, forest clearance for developmental projects and public's concern for their health and the health of their natural environment. This led to a unifying movement known as 'Environmentalism. We are aware of Chandi Prasad Misra movement ably supported by Sunderlal Bahuguna (Chipko movement) to protect forests in Uttarakhand by hugging the trees and not allowing forest contractors to cut the trees, the 'Narmada Bacchao' movement, 'The Gandhamardan movement' not allow mining of bauxite ore by Balco Company etc.These are examples of Environmentalism.

Environment houses resources and is an enabler to provide livelihood support sustainably if protected and managed scientifically. Environmental education was thus born of the realization that solving complex local, regional and global environmental problems cannot be accomplished by politicians and experts alone, but requires "the support and active participation of an informed public in their various roles as consumers of resources, employers, business and community leaders and voters.

The figures 4A-D, and 5 are natural diverse ecosystems but they have common characteristics of multispecies, (type of species are different), stable, having complex food web, largely nonpolluting, solar energy dependent ecosystems. They provide resources, especially Bioresources to human beings in sustainable manner. Manmade ecosystems such as agriculture system, aquaculture system, industrial system, and urban systems are energy intensive, pollution generating but highly productive and fragile. These are usually monospecies/ no species systems. They are not sustainable unless scientific and resource management are practiced.

These systems require huge amount of non-solar (fossil fuels) energy, resources and labor for their running. These systems, however provide jobs and productive, although they generate pollutants that create health hazards.

Figure 4A. A Forest Ecosystem in Odisha India (provide with diversified niches to biodiversity, food, fodder, ayurved medicine, human habitation material etc to Man).

Figure 4B. Canadian prairie Grassland Ecosystem (provide fodder, grazing land etc to Man and Cattle).

Figure 4C.Chilika-Nalabana ecosystem.

Figure 4D. A natural mangrove ecosystem (rich in biodiversity).

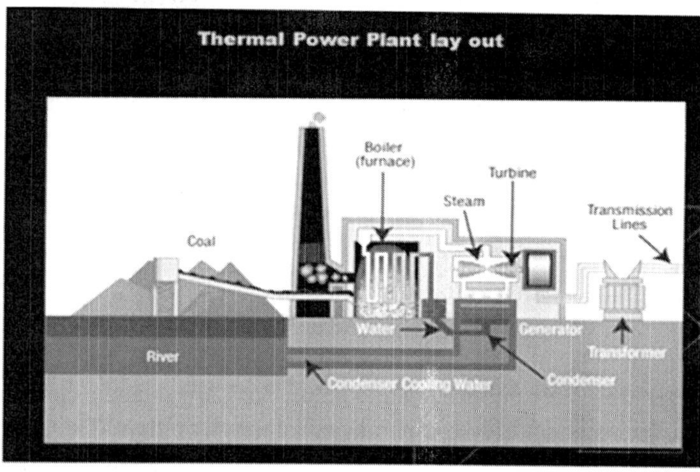

Figure 5. A man-made system, heavy resource input (mainly water and coal), cause huge pollution.

A definition of "Environmental Education" first appeared in The Journal of Environmental Education in 1969, authored by William B. Stapp, who later went on to become the first Director of Environmental Education for UNESCO. Simultaneously, another first articles about environmental education as a new movement also appeared in the Phi Delta Kappa in 1969 authored by James A. Swan.

Ultimately, the first Earth Day on April 22, 1970 paved the way for the modern environmental education movement. Later that same year, National Environmental Education Act was enacted in USA, which was intended to incorporate environmental education in Higher Secondary Schools. In 1971, the National Association for Environmental Education (now known as the North American Association for Environmental Education) was established to improve environmental literacy by promoting environmental education programs.

Internationally, environmental education gained recognition when the UN Conference on the Human Environment held in Stockholm, Sweden, from June 5 to 16, 1972, declared environmental education must be used as a tool to address global environmental problems. The United Nations Education Scientific and Cultural Organization (UNESCO), and United Nations Environment Programme (UNEP, 2011) created three major declarations that have guided the course of environmental education. In India in late 1990s, the Supreme Court ordered to introduce Environmental Studies in + 3 level in undergraduate studies.

Stockholm Declaration: June 5–16, 1972 - The Declaration of the United Nations Conference on the Human Environment. The document was made up of 7 proclamations and 26 principles "to inspire and guide the peoples of the world in the preservation and enhancement of the human environment."

Belgrade Charter: October 13–22, 1975 - The Belgrade Charter was the outcome of the International Workshop on Environmental Education held in Belgrade, Yugoslavia (now Serbia). The Belgrade Charter was built upon the Stockholm Declaration and detailed the objectives, and guiding principles of environmental education programs, for schools and for general public. (Wikipedia .com)

Tbilisi Declaration: October 14–26, 1977 - The Declaration "noted important role of environmental education in the preservation and improvement of the world's environment, as well as in the sound and balanced development of the world's communities." The Tbilisi Declaration updated 'The Stockholm Declaration' and 'The Belgrade Charter' by including new goals, objectives, characteristics, and guiding principles of environmental education (.Wikipedia .com) (Wikipedia .com)

The components of environmental education are:

1. Awareness and sensitivity to the environment and environmental challenges,
2. Knowledge and understanding of the environment, Ecosystems, and environmental challenges,
3. Attitudes of concern for the environment and motivation to improve or maintain environmental quality,
4. Skills to identify and help resolve environmental challenges,
5. Participation in activities that lead to the resolution of environmental challenges

Environmental education does not advocate a particular viewpoint or course of action. Environmental education teaches individuals how to weigh various sides of an environmental issue through critical thinking and it enables them in solving their problems and decision-making skills. Environmental education increases public awareness and knowledge of environmental issues, provides facts about environmental issues and enhances individual problem solving capability, Environmental education (EE) involves organized efforts to teach how natural environments function, and particularly, how human beings can manage ecosystems to live sustainably. Environmental Science and Education is a multi-disciplinary field integrating disciplines such as biology, especially ecology, earth system science (geology, geography), chemistry, physics, atmospheric science, and mathematics. However, environmental education must include efforts to educate the public and print media, websites, media campaigns, et solutions to these issues. The UNESCO) states that EE is vital in imparting an inherent respect for nature amongst society and in enhancing public awareness. UNESCO emphasizes the role of EE in safeguarding future global developments of quality life through the protection of the environment, eradication of poverty, minimization of inequalities and insurance of sustainable development (UNESCO, 2014a, b).

Focus on Environmental education: The focuses are on:

1. Engaging with citizens of all ages to;
2. Think critically, ethically, and creatively when evaluating environmental issues;
3. Make educated judgments about those environmental issues;

4. Develop skills and a commitment to act independently and collectively to sustain and enhance the environment; and,
5. To enhance their appreciation of the environment; resulting in positive environmental behavioural change (Bamberg & Moeser, 2007) ;

An Environmental education curriculum to explore the human-nature relationship

Environmental Education encourages the discovery and understanding of the Earth's natural systems (Forests, Grasslands, Lakes, Geo-systems, and Bio-systems etc) and the human role in those systems. Environmental educators teach about the interrelationships among all living things and with surroundings. They have a solid comprehension of ecological concepts, (on ecosystem productivity, energy flow, consumption patterns, matter circulation, balance of nature etc) and an understanding of environmental history and the ecological effects that humans have had on the Earth and the homeostatic mechanisms. Students pursuing this emphasis explore the literature and philosophy concerning the human–nature relationship.

Environmental educators must have ecology, environmental science, economic, political, and environmental law background in order to teach about relationships among local communities, technological society, and the local, regional and global environment. A foundation in the field of education with an understanding of learning theories, curriculum design, and experiential and field study education, appreciation of the beauty of the natural world provides them with the necessary knowledge and skills to develop their teaching and to demonstrate their competence. (Tanner, 1974).This is a big challenge.

Benefits of Environmental Education: Few important benefits are described below.

(i) Imagination and enthusiasm are heightened in Youth

EE sparks the imagination and unlocks creativity among youth. When EE is integrated into the curriculum, students are more enthusiastic and engaged in learning.

(ii) Learning transcends the classroom to the Real World

Not only does EE offer opportunities for experiential learning outside of the classroom, it enables students to make connections and apply their learning in the real world. EE helps learners see the interconnectedness of social, ecological, economic, cultural, and political issues.

(iii) Skill Development and Enhancement

EE encourages students to research, investigate how and why things happen, and make their own decisions about complex environmental issues. By developing and enhancing critical and creative thinking skills, EE helps foster a new generation of informed consumers, workers, as well as policy or decision makers. In the evolution of Environment study, the discipline has developed into two broad categories i/e. (i) Environmental Science consisting of rigorous scientific analysis of theory, field scientific studies involving monitoring, and environmental management etc; keeping human interest and sustainability in mind; (ii) Environmental studies involving society, creating awareness and sensitizing people, especially school students and public at large.

(iv) Tolerance and understanding are inculcated

EE encourages students to investigate varying sides of issues to understand the full picture. It promotes tolerance of different points of view and different cultures.

(iv) Interdisciplinary Learning

By incorporating EE practices into the curriculum, teachers can integrate science, math, language arts, history, and more into one rich lesson or activity, and still satisfy numerous state and national academic standards in all subject areas. Taking a class outside or bringing nature indoors provides an excellent backdrop or context for interdisciplinary learning.

(vi) Appreciating Nature's Beauty and playing for fun

By exposing students to nature and allowing them to learn and play outside, EE fosters sensitivity, appreciation, and respect for the environment. It combats "nature deficit disorder" ... and it's FUN!

(vii) Healthy lifestyles are encouraged

EE gets students outside and active, and helps address some of the health issues we are seeing in children today, such as obesity, attention deficit disorders, and depression. Good nutrition is often emphasized through EE and stress is reduced due to increased time spent in nature.

(viii) Communities are strengthened

EE promotes a sense of place and connection through community involvement. When students decide to learn more or take action to improve their environment, they reach out to community experts, donors, volunteers, and local facilities to help bring the community together to understand and address environmental issues impacting their neighborhood.

(ix) Responsible action for betterment of the environment

EE helps students understand how their decisions and actions affect the environment, builds knowledge and skills necessary to address complex environmental issues, as well as ways we can take action to keep our environment healthy and sustainable for the future. Service-learning programs offered by PLT and other EE organizations provide students and teachers with support through grants and other resources for action projects.

(x) Empowerment of Students and Teachers

EE promotes active learning, citizenship, and student leadership. It empowers youth to share their voice and make a difference at their school and in their communities. EE helps teachers build their own environmental knowledge and teaching skills.

Through an interdisciplinary curriculum that covers topics such as ecosystem analysis, biodiversity, & other resources conservation, environmental thought, environmental debates, chemistry of pollution, its effects on biodiversity and man, and its abatement, climate change and environmental politics, one will gain an understanding of the theories, policies, and ethics that have shaped public discourse and understanding of our environment; examine national and local models of environmental education; analyze their relationship to contemporary issues of environmental literacy; and develop competencies in planning environmental education programs. In Indian context folk song, folk dances, Jatra, Daskathia, Pala, etc can create environmental awareness in people, especially village people.

During the last few decades development of environmental studies has grown in two interrelated aspects such as (i) the scientific and technological aspects of Environmental Science; and (ii) softer way to sensitizing people in various ways; and to abate environmental degradation by creating legal instruments. Environmental laws in India (The Wildlife Act-1972, The Forest Act-1980, The Water (Pollution prevention and Control) -Act, 1974, The Air (Pollution prevention and Pollution Control) Act-1981, The EP (Environmental Protection) -Act-1986 (enacted after Bhopal Gas Tragedy) and the interpretation of the Supreme court on Article-21 (Right to Life) of the constitution, the constitution of Appellate Authority and the Green Tribunal have now become integral part of legal education. Besides, all developmental projects before implementation are to assess the impacts the project (Environmental Impact Assessment) would create if implemented and to look into the environmental management plan for sustainable development.

References

Bamberg, S.; Moeser, G. (2007). "Twenty years after Hines, Hungerford, and Tomera: A new meta-analysis of psycho-social determinants of pro-environmental behaviour". *Journal of Environmental Psychology. 27 (1) : 14–25.*

Buhler, G.1886, The Laws and Manu.In'The Sacred Books of the East' (edited by F.Max Muller, Oxford in Clarinda Press, London.

Dash, M.C. and S.P.Dash.2009. Fundamentals of Ecology, McGraw-Hill Publications, New Delhi, 562 pages, Earthday.net

Padhy, S.,, S.K.Dash and Ratnaprave Mohapatra., 2006.Environmental Laws of Manu: A Concise Review, Journal of Human Ecology, 19 (1) :1-12.

Tanner, R.T. 1974. "Ecology, environment and education, " Lincoln, NE, Professional Educators Publications.

Swan, J.A. (1969). "The challenge of environmental education". Phi Delta Kappan. 51: 26–28. Stapp, W.B.; et al. (1969). "The Concept of Environmental Education" *(PDF). The Journal of Environmental Education. 1 (1) :* 30–31.

The Belgrade Charter, Adopted by the UNESCO-UNEP International Environmental Workshop, October 13–22, 1975.

UNESCO (United Nations Educational, Scientific and Cultural Organization)., 2014a. Ecological Sciences for Sustainable Development. [online] Available at: <http://www.unesco.org/new/en/natural-sciences/environment/ecological-sciences/capacity-building-and-partnerships/educational-materials/>

UNESCO (United Nations Educational, Scientific and Cultural Organization)., 2014b. Shaping the Future We Want: UN Decade of Education for Sustainable Development. [pdf] Paris: UNESCO. Available at: < http://unesdoc.unesco.org/images/0023/002301/230171e.pdf>

Wikipedia.com: the free encyclopedia on Environmental Education www.rdasia.com

2

Ecosystem Analysis: Concept of Ecosystem and Ecosystem Diversity

Introduction

All forms of life owe their existence to the mother Earth and to the maintenance of ecological order (ecological balance) in nature. This balance is maintained by:

(i) Rich bio-diversity, (Microbes, Plants, Animals, & Man), (ii) Interdependence among the different life forms, (iii) Sustained productivity and food chain relationships, (iv) Maintenance of natural processes of nutrient cycling, energy, flow, homeostatic mechanisms and evolutionary process, (v) Adapting potentiality of the living forms to the changing environment.

The biosphere consisting part of hydrosphere (water), pedosphere (soil), atmosphere (air) and the living forms (bio-diversity) have arisen in the evolutionary process. The living microorganisms, plants and animals have specific requirements for existence, continuance and perpetuation.

All the living forms including man get their requirements of gases, other matter, and energy from the environment where they live. Ecology, a word derived from the Greek word Oikos, meaning house, is concerned with the biology of organisms, populations, communities etc. and their functions in natural surroundings. In another words, ecology deals with the understanding of the structure and function of nature; this includes the living world including man and the nonliving substances, which are constantly taken by the living organisms from their surroundings to keep them alive. As a scientific discipline, ecology is about 137 years old (the German Zoologist coined the word Ecology in 1876), although man is aware of its association with nature & nonliving matter since the age of Vedas.

The practical importance of ecosystem analysis results from the presence of man in the ecosystem. Man's evolution, its dependence on environment, his relationship with nature and the development of his position from one of struggle against natural forces to the recent influence and use and abuse of it are included in ecology and ecosystem analysis.

Pre-agricultural human society, agriculture- based society, industrialization, urbanization, the atomic age and the electronic age, and the global warming and climate change have brought the science of ecology to the forefront of human knowledge.

Ecosystem Concept

The living microbes, plants and animals of a habitat & their nonliving environment function together as these are connected to each other through food chain, matter circulation, energy flow etc. as one unit, which is called as an Ecological system or Ecosystem. Man is part of the ecosystem. This concept is holistic as all component parts of the ecosystem function together and has functionally obligatory relationship. A natural pond, Lake, Bay, Ocean, Forest, Grass lands etc. are considered as a natural ecosystem.

The component parts are :

(i) the abiotic (nonliving) substances found in the atmosphere, water, soil and factors like temperature, relative humidity, sunlight, rainfall etc. associated with the environment. This component changes with time and affects the activities of living processes,

(ii) green plants called producers or energy transducers, which convert solar energy into chemical energy stored in plant tissue and this energy becomes the source of energy for other living organisms including man in the ecosystem,

(iii) Consumers like plant grazers (grasshoppers, cattle, etc.), flesh eating animals like frogs, snakes, some birds, and many mammals like tigers and omnivores like man form another component part

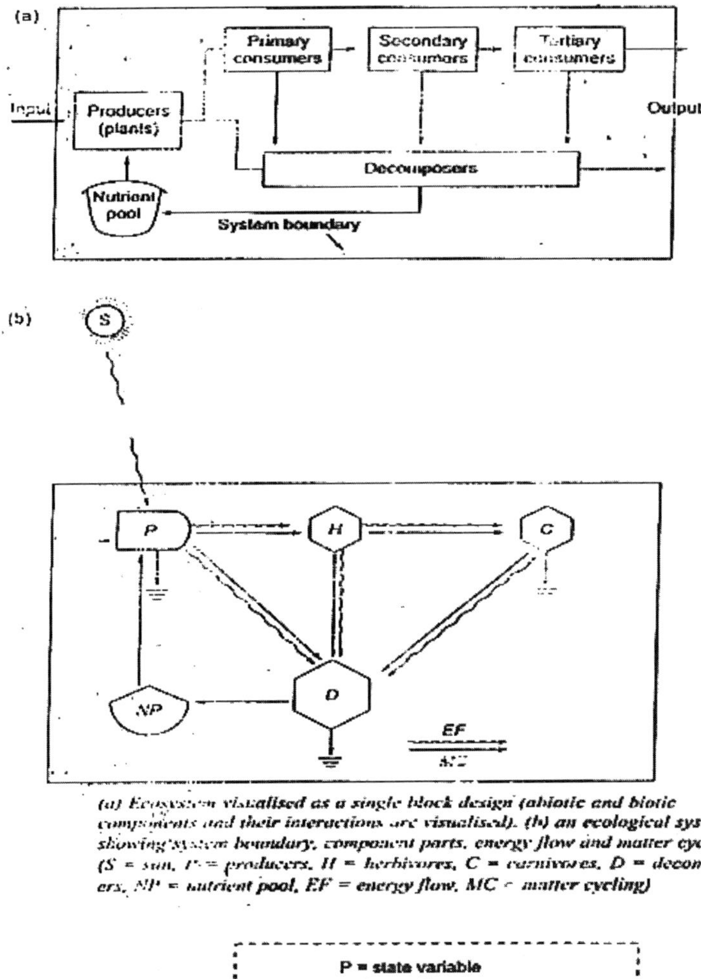

(a) Ecosystem visualised as a single block design (abiotic and biotic components and their interactions are visualised). (b) an ecological system showing system boundary, component parts, energy flow and matter cycling (S = sun, P = producers, H = herbivores, C = carnivores, D = decomposers, NP = nutrient pool, EF = energy flow, MC = matter cycling)

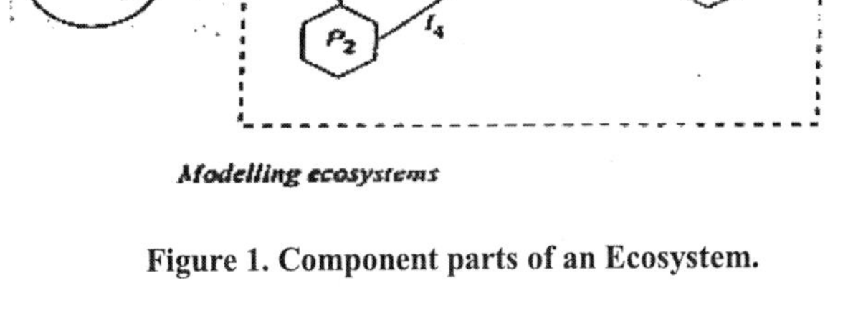

Modelling ecosystems

Figure 1. Component parts of an Ecosystem.

of this system. Plant grazers are called herbivores & animal flesh eaters are called carnivores, which may be classified under different categories like Carnivore-1, Carnivore-2 so on,

(iv) Decomposers form another component. They are detritus eating organisms like bacteria, fungi and some animals like Earthworms.

The green plants are the autotrophs. The consumers & decomposers are the heterotrophs. The green plants, the consumers such as herbivores and carnivores, the detritus feeders such as bacteria, fungi, and soil invertebrate animals form the biodiversity of an ecosystem. Man is also part of the ecosystem. Photographs of natural ecosystems and list of biodiversity found in ecosystems are given below.

Structure in the ecosystem level of integration includes diversity and relative abundance of species, number, distribution and the biomass of various trophic groups, and the sizes, source and locations of energy and nutrient pools, and the conditions of the physical environment. Function in such system includes rates of energy and nutrient transfer and the operation of homeostatic mechanisms to maintain the structural stability of the system.

None of these levels is more important than another with regard to teaching and research in ecology.

Students at the school level should be given field & practical training on ecological concepts,, to develop interest in field studies and to design and conduct simple experiments, and use quantitative methods for analysis and interpretation of results.

Solar energy input can be a forcing function, the producer biomass, herbivore and carnivore biomass can be state variables, the matter/ energy flow can be designated as f1, f2 and so on, and I as interaction between the state variables. These variables can be quantified ecosystem wise and systems can be modeled.

The natural ecosystems of the world include Tundra, Desert, Lakes, Rivers, Ponds, Oceans, Forests (Tropical, Subtropical, Mangroves, Temperate, Alpine, etc), and Grasslands in different latitudes and Altitudes etc (Figures 2—7). These ecosystems may vary with respect to species composition at producer, consumer and decomposer levels but they do similar functions in every ecosystem. Producers' transduct solar energy and produce carbohydrates, consumers (herbivores and carnivores) do their functions of grazing and predating, decomposers decompose organic matter and in the process materials are released for use of the biodiversity & human beings. Photographs of a subtropical forest ecosystem, a natural grassland, a temperate region lake ecosystem, a mangrove ecosystem, are given below.

Figure 2.A Natural Forest Ecosystem

Figure 3.A Natural Grassland Ecosystem

Figure 4. A Natural Temperate Region Cold Water Lake Ecosystem.

Figure 5. A Natural sub-tropical Mangrove Ecosystem in Odisha, India.

Figure 6. Mangrove Ecosystem showing the stilt roots of mangroves.

Figure 7. A Natural ecosystem with forest, water fall and a fresh water pond from Odisha.

India is a mega biodiversity country having about 8-9% of world's biodiversity although land area is only 2.4% of the planet. The Himalayan ecosystems & the Western Ghats ecosystems are international biological hot spots. There are 26 Ramsar Sites in India. The Chilika Lake in Odisha was the first Ramsar site declared in Ramsar Convention. There are 18 mega biodiversity countries in the world.

The beauty of ecology lies in understanding the natural order, which keeps so many living organisms in a dynamic coexistence.

This raises many fundamental questions in human mind.

What are the fundamental features of these natural systems?

How the nutrients circulate and energy flow in natural systems?

How efficient are the natural systems in comparison to manmade systems such as paddy fields, wheat fields, aquaculture ponds, etc?

Is there something called carrying capacity of an ecosystem?

How can this knowledge be utilized to increase food, fodder and fuel production, to protect environment (landscape) from erosion, pollution etc. and manage the resources?

The functioning of natural systems is based on certain basic principles and understanding of these principles will help man to solve many of its own problems.

In the modern concept, Ecological principles include: First Principle: "Nothing comes from nowhere, everything comes from somewhere". The Second Principle: Nothing ever goes away, everything goes somewhere .The science of ecology has three aspects: (i) Theory; (ii) Practical; (iii) Socio-economic-cultural & Aesthetics.

The structure of an ecosystem refers to the diversity, distribution, numbers and biomass of living organisms in the ecosystem, the qualities and distribution of organic nutrients and the physicochemical conditions characterizing the system. Ecosystem function/ services refers to the energy flow and its rates, nutrient cycling, biogeochemical cycles such as water cycle, mineral cycles, oxygen and carbon dioxide and other gaseous cycles, and the homeostatic mechanisms that determine the structure and stability of the system.

Ecosystem Enumeration

Organisms, populations, and the Communities are three important components of any natural ecosystem..

The organism level of the integration deals with the relationship of the individual organism and its immediate biotic and abiotic environment. The structural and functional aspects of these systems correspond to the morphological, physiological, biochemical and behavioral adaptations of the organism and the characteristics of the environment, to which the organism is exposed and the interactions related to this exposure (Cox 1970). The specialized fields of the functional morphology, ecophysiology and behavioral ecology direct much of their attention to relationships at organismal level.

The population level of integration deals with the density, distribution and dispersion pattern, and sex and age structure of the populations, together with the descriptions of the biotic and abiotic characteristics of the environment. Function in such systems is characterized by natality and mortality, rates of transfer of nutrients and energy along trophic levels, and patterns of modifications of environmental conditions and the operation of mechanisms of population regulation.

The Community level: A community consists of populations in a specified space.

Levels of different organizations are given below: Conceptual hierchical model of matter and living systems at micro and macro level is given below.. Autecology deals with Organisms and Population, and Synecology deals with Community ecology/landscape ecology etc.

The levels of study of the Universe/Solar system is given in Figure 8.

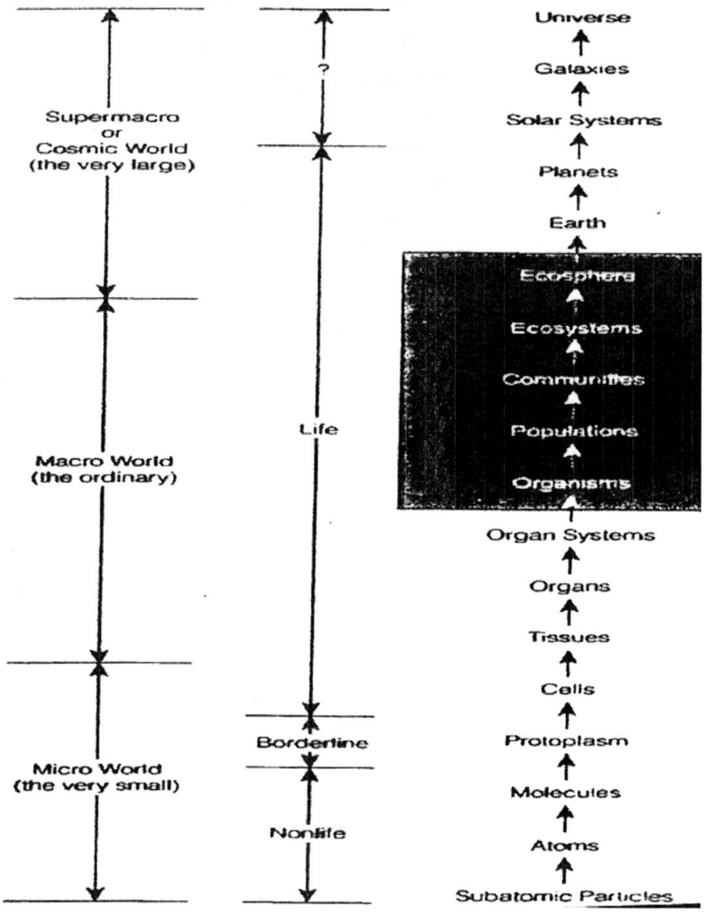

Figure 8. Levels of Study.

Ecosystem includes atmosphere (air), hydrosphere (water), and lithosphere (soil) (.Figure 9-11).

The Stratification of the Atmosphere, and the major biological systems, that operate in soil and stratification of Oceanic environment are given in Figures 9-11These systems operate in the Biosphere, the zone where land, water, atmosphere and biodiversity interact and the biosphere includes the diverse ecosystems.

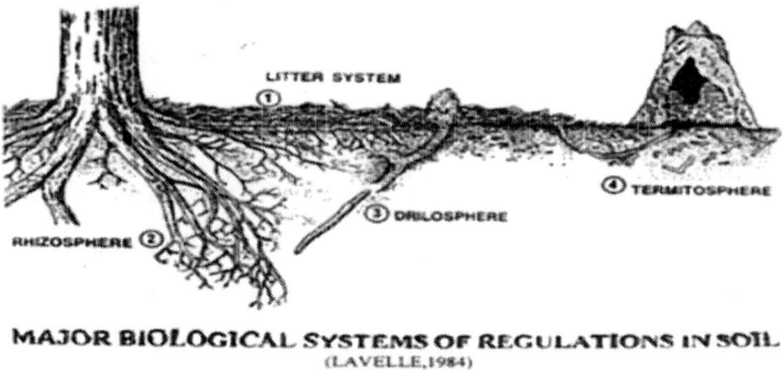

Figure 9.

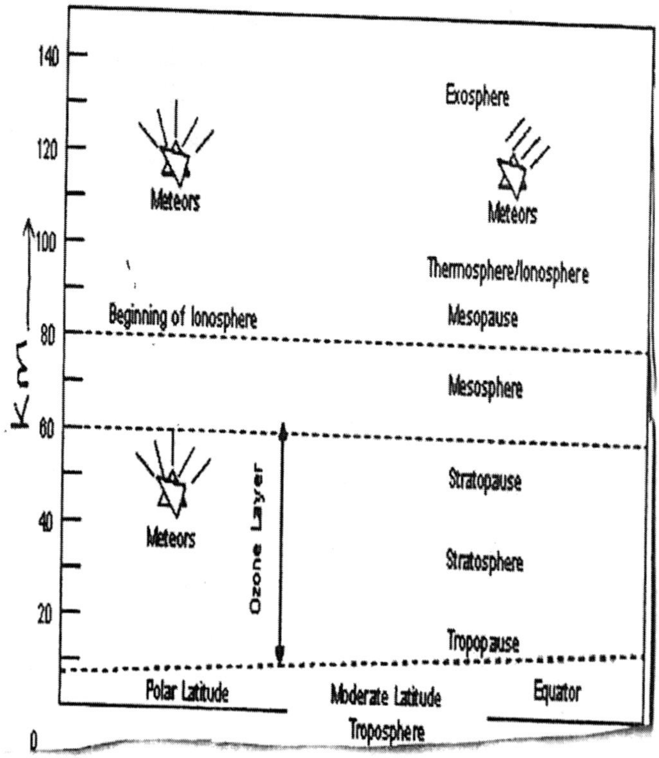

Figure 10. The major Biological systems operate in Soil.

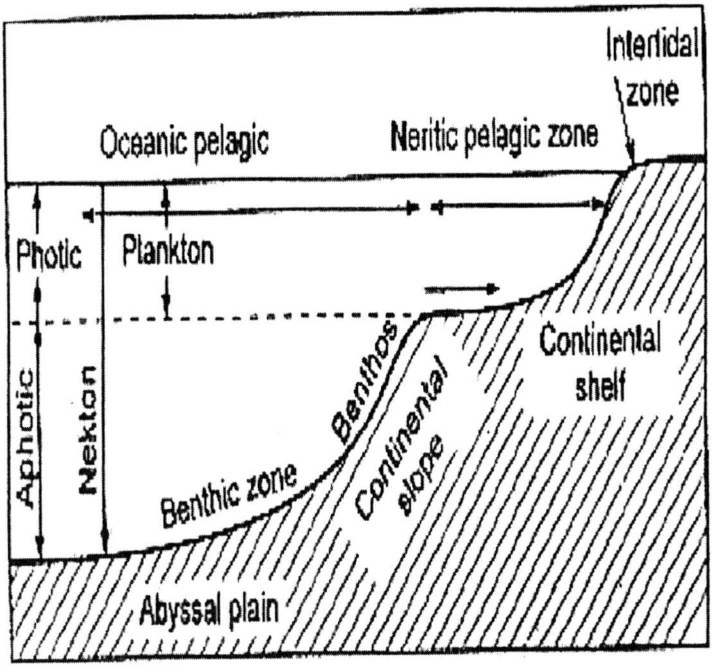

Figure 11. The Oceanic Environment.

Comparison of Natural Ecosystem and Man Made Ecosystems

The natural systems are largely pollution free and less energy intensive systems. These systems provide with food, fodder, and all resources to man. Agro ecosystems, Aquaculture systems, Industrial systems, urban systems etc are man-made systems. They generate pollutants that create health hazards.

These man-made systems require huge amount of energy, natural resources and labor for their running. These systems however provide jobs and other requirements to growing population.

Functional Attributes of the Ecosystem can be grouped as:

Bio-diversity (2 to 30 million species) ;taxonomical, ecological and genetically and maintenance of stability.

Productivity (both Plant (primary) and Animal (secondary), Food web relationships, Matter (Nutrient, gases and water) cycling between the component parts of the ecosystem, Energy flow processes.

Feedback mechanisms and Homeostatic processes, which keep the ecosystem stable,

Development & evolution of ecosystems. The ecosystem services can be referred to the functional attributes.

Advantage of Ecosystem Concept

Adoption of ecosystem concept and identifying component parts and functional attributes has helped man to understand nature in a holistic manner &manage the natural and man made ecosystems to its own advantage..

Ecosystems can be modeled, and inputs, outputs, etc. can be quantified and predictions under defined conditions can be made.

Modeling has become essential for ecosystem management.

Energy Flow Based Classification

On the basis of energy flow ecosystems can be classified into different categories like:

(i) Unsubsidized, solar powered natural ecosystems,

(ii) Nature subsidized natural solar powered ecosystem,

(iii) Man subsidized but solar powered ecosystems and

(iv) Man made fossil fuel powered urban-industrial ecosystems.

Natural ecosystems have provided food, fodder, medicine, clothes etc. to man since time immemorial but now they are under intensive pressure due to many anthropogenic activities.

The natural systems are largely pollution free and less energy intensive systems.

Manmade ecosystems (Agriculture, Aquaculture etc) are energy intensive, pollution generating but highly productive

Disturbance in Ecosystem & Balance of Nature

Any disturbance in the system whether natural or man-made brings unpredictable changes in the system and may disrupt links and restoration involves money, time etc.

Disturbance (natural-earthquake, volcanic eruption, cyclone etc) in natural system may be unpredictable. Human disturbance (change in land use, industrialization, pollutant generation, man created fire etc) create huge problems.

There is need of Environmental management: the system may be natural or man-made

Another example can be that the material cycles like Oxygen, Carbon dioxide, sulphur, phosphorus, calcium, magnesium etc. operate in nature due to interaction of biodiversity and abiotic components.

Plants utilize carbon dioxide in food production during day time and release oxygen. Animals & man utilize oxygen for metabolism and release carbon dioxide. A balance is maintained.

Balance is disturbed:

Because of anthropogenic activities like deforestation, use of automobiles in very large scale, heavy industrialization, urbanization etc., more carbon dioxide (green house gas) is produced than utilized by plants. More carbon dioxide in air absorbs more heat and makes the planet (troposphere) warmer. The balance of nature is not maintained.

Man's interference disrupts ecosystem processes as every ecosystem has a limit to its carrying capacity for normal functioning. Maintenance of ecosystem productivity, number (biomass) of herbivores, carnivores etc is very important and man interfers and breaks the balance.

Environment in India

The total geographical area is 328 million hectare (3.28 million sq.km.).

The total forest area is 0.76 million sq. km. (76 million hectare), which forms 23.2% of the total geographical area.

India has a land frontier of 15200 km., seacoast runs to about 6100 km.

Biological diversity is very rich, having about 48, 000 plant species including 15000 flowering plants and many of them are endemic and 81000 species of animals.

Biodiversity will play crucial role in future for survival of mankind, if it is conserved with utmost care. Each species has a life cycle and therefore it is maintainable.

At the time of independence from British rule, agriculture was stagnant, industrialization poor and lopsided, transport communication under developed and social security scheme non-existent, death rate high, poverty rampant, unemployment wide spread.

Economic development plans were set within a framework of parliamentary democracy, guarantying universal franchise.

First 5 year plan started in 1951. Rapid economic development created enormous pressure on natural resources, forests became degraded, industrialization without pollution control measures lead to water, air and land pollution.

From 1950-1970 environmental issues were completely overlooked. After the Stockholm conference the Central Pollution Control Board and the State Pollution Control Boards were established in India to look into the pollution prevention and control problems.

Eco-city ecosystem Concept

Ecocity Concept: On a separate platform at Rio in 1992, experts emphasized on local issues in urban areas. These issues are:

Population migration, housing, poor sanitation, transport & traffic congestion, Solid waste management, sewerage system management, accidents, noise, plantations, etc which have direct impact on human life style and health.

The issues are air, water, soil & hazardous waste pollution and management of urban centers. In view of these facts, the eco-city (man-made urban ecosystem) concept developed.

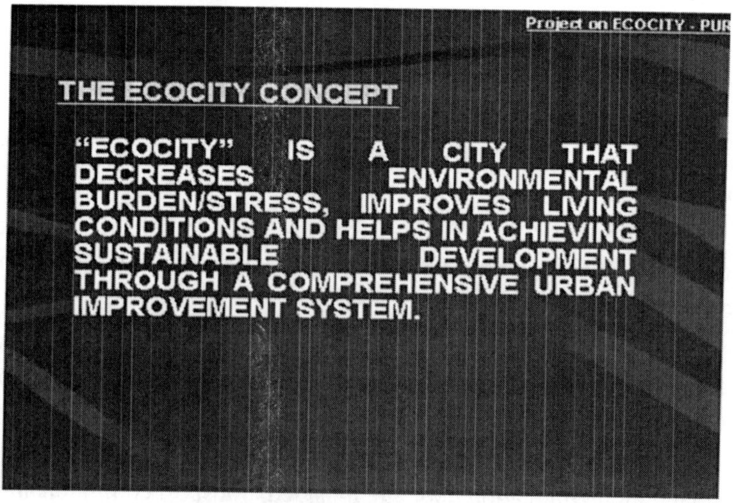

Figure 12. Eco-city-Ecosystem Concept.

We must conserve our natural ecosystems for our survival & continuance on this planet.

Environmental laws have been enacted to maintain Ecological Order.

The laws are: Water Act, (Water Pollution Prevention & Control) 1974, Air Act, (Air Pollution Prevention & Control) 1981, Environment (Protection) Act, 1986 followed by many notifications to management of natural systems and control of Environmental degradation. The Forest Act, 1980, The Mining Act, 1927, amended recently. Many other Acts & notifications to prevent & control environmental degradation. Creation of massive awareness among the people on the subject of ecosystem conservation is important to save the future of mankind. We should respect it for our survival & continuance on the planet. The Article 21 of the Constitution of India (right to life) and its interpretation by the Supreme Court of India is very welcome to protect life in this planet.

References

Dash, M.C. and S.P.Dash. 2009. Fundamentals of Ecology, McGraw-Hill Publication, New Delhi, 565 pages.

Besides, a number of papers and books were consulted to write this article.

Part I : Biodiversity

3

What is Biodiversity?

Introduction

The UNO Conference on "Man & Environment" at Stockholm on 5th June 1972 and the "Earth Summit" in Rio de Janeiro (Brazil) in 1992 and Johannesburg (South Africa) Conference, 2002 renewed the consciousness among the people in general and the leaders around the world to address the problems of the environment along with Economic development. It was realized that sustainable development is the key for long term security on earth. There is a growing concern for devising and implementing the concepts of sustainability in to economic practice.

The Vedic hymn says "Sustainable development cannot be achieved unless peace prevails in the environment". Modern concept of sustainability has to look into efficient and effective maintenance of (i) Biological order (conservation of Biodiversity and utilization in a sustainable manner), (ii) Economic order, (iii) Social order, (iv) Scientific & Technological input and (v) Environmental order, which includes prevention and control of Environmental pollution, biodiversity conservation, polluter pays principle, principles of cooperation between the state and citizens. Caring of environment has (i) Existential function (mankind must exist) enjoy life worthy of human dignity, (ii) Social function, (iii) Aesthetic function. The task is enormous and the infrastructure facilities available are meager. Hence it is a challenged and requires full commitment to have an ever-clean environment for sustainable development.

However, the Wild Life (Protection) Act, 1972, the Water Act, 1974, the Forest Conservation Act, 1980, the Air Act, 1981, the E.P. Act, 1986, the Biodiversity Act, 2002-03, the Seed Act, 2005 and the Patent Act, 2005 and all the regulations and rules and decisions of the Hon'ble Apex and High Court in India are the legal instruments to conserve Nature and to achieve sustainable society. Definition as per Earth Summit, 1992:

The Agenda-21 and Article-2 of Rio Conference (Earth Summit, 1992) defines "Biodiversity as variability among living organisms in all sources including interalia, terrestrial, marine and other aquatic ecosystems and the ecological complexes of which they are part". This includes diversity within species, between species and of ecosystems. The important thing is recognition of ecological patterns, complexes and ecosystem diversity. This is important when we think of their (i) sustainable use, (ii) conservation, (iii) fair and equitable sharing of the benefits arising out of utilizing of genetical resources and access and rights over these resources leading to technology transfer.

Planet's biodiversity is organized into three levels namely (i) ecosystems and ecological patterns, such as rain forests, lakes, coral reefs, Himalayan ecosystem, Western ghats and Eastern ghats, Chilika lagoon, Sunderban, Bhittarkanika mangroves etc (ii) species level that comprise about 2 million species identified by biologists so far and may be another few millions yet to be identified, (includes Inter and Intraspecific variations). ((iii) genes (genetic diversity) which make up the heredity of each species and is now considered the basic important tool for Genetic Engineering and creation of transgenic plants and animals and tool for biotechnology research.

Biodiversity also involves (a) taxonomical, (b) ecological, (c) adaptational, (d) genetical and (e) biochemical diversity etc. Hence Biodiversity involves: Genes – Species – Communities – Ecosystems – Ecological patterns. Biologists estimate that atleast half of the species occur in tropical rain forests. A tropical rain forest may possess 425-820 kinds of woody trees in one hectare of land in comparison to

5-50 species of woody tree species in middle or temperate region. One hectare of tropical forest in Malaysia has 820 species of woody plants.

Foundations: Biodiversity highlights the following and forms foundations for the following ecosystem processes and human need. Scientists are of the opinion that

(i) Few million species exist in the world and only 10% have been studied from utilitarian viewpoint.

(ii) Interdependence of organisms happens naturally,

(iii) Food chain relationships and multi channel food web occur in nature,

(iv) Balance of Nature and maintenance of Ecological order depend on biodiversity and different processes,

(v) Sustained productivity by the biodiversity is required for continuance and maintenance of ecosystems and for ecosystem stability.

(vi) Multi species patterns generate stable ecosystems.

(vii) Foundation for Agriculture, Industry, especially pharmaceutical and Ayurved industries and Civilization (Food, Fodder, Fuel, Fibre, and Medicine) is dependent on the biodiversity.

(viii) Biodiversity is the life supporting system.

Today's bio-diversity is the result of evolution from a single celled organism some 3.5 billion years ago. This unitary origin explains why all organisms share the same basic chemistry: DNA is always the storage molecule of genetic information and the complex process of protein biosynthesis is virtually the same in all organisms. Separate species evolved when mutations occurred and between relatives no longer allowed for interbreeding, for instance after geographic or reproductive separation .Planet's biodiversity is organized into three levels (genes—species—ecosystem level) as described above.

In Ecological Terms Biodiversity are three types namely

(i) Alpha diversity- quantitative measure of species richness in a given habitat/ecosystem;

(ii) Beta diversity- Quantitative measure of relative change in species composition from one ecosystem to another;

(iii) Gamma diversity- total richness of species and its turnover in a geographical region/area (landscape) consisting of different habitats.

The rich biological diversity of India is largely due to prevalence of wide range soil, climatic conditions and habitats and it is reflected with rich flora and fauna and diversified ecosystems. Biodiversity of, land scapes, ecosystems etc is usually quantified using Shannon and Wiener index and some other indices.

The total geographical area of India is about 328.73 million ha having a forest cover of about 78.38 million ha. (~23.84%). India has about 50, 000 plant species with 5, 000 endemic species and about 84, 000 animal species, so far identified. India had earlier some 30, 17, 009 sq. km. of wilderness and about 80% of it have vanished due to human activities without replacement within the last 70 years or so.

Biodiversity and Ecosystem Functioning:

The relationships between biodiversity, productivity, stability of the ecosystems and some other parameters have been studies and these

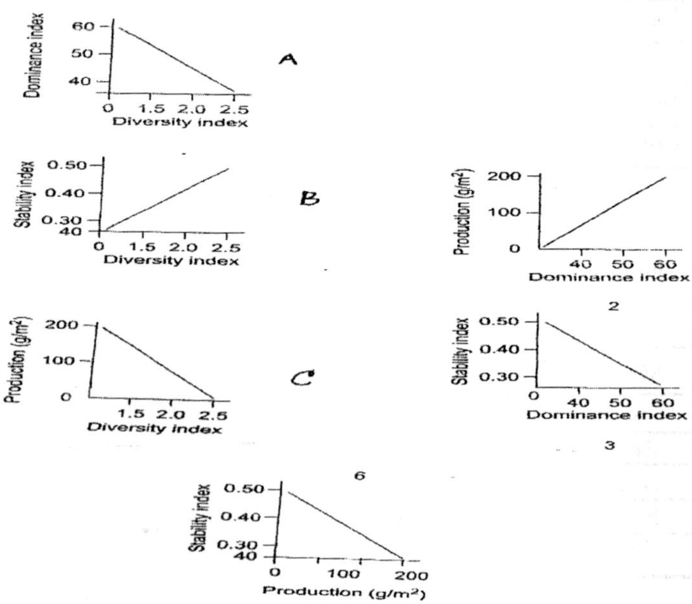

Relationship of species diversity, dominance and stability in grassland ecosystems (based on McNaughton, 1967, Krishnamurthy, 1978 and others, see text).

Biodiversity and Ecosystem Functions.

relationships are described below. Biodiversity of an ecosystem can be quantified and index can be developed using many type of indices. The commonly used diversity indices are the Shannon-Wiener index and Simpson index. (Dash and Dash, 2009, Fundamentals of Ecology, McGraw-Hill Publ., New Delhi, 3rd edition).

These graphs show the relationships of different ecosystem functions with Diversity index

Biodiversity Enumeration:

Table 1, 2 and 3 gives the number of living species of all organisms recoded by scientists.

Table 1. Number of living species of all organisms (Wilson, 1992).

Organisms	Number of species
Insects	751,000
Other Animals	281,000
Higher Plants	248,000
Fungi	6,900
Algae	26,900
Protozoa	30,800
Bacteria and similar form	4,800
Viruses	1,000

Table 2. Recent Estimated of species of biodiversity in the world.

Group	Number of species in India (SI)	Number of species in the world (SW)	SI/SW (%)
Mammals	372	4,629	08.03
Birds	1,224	9,702	12.6
Reptiles	408	6,550	06.2
Amphibians	197	4,522	04.4
Fishes	2,546	21,730	11.7
Flowering Plants	15,000	2, 50,000	06.0

Table 3. Biodiversity in India

Floristic (Flora) Groups	No. of spy.	Faunistic (Animals) Group	No. of spp.
• Angiosperms	>15000	Mammals	> 372
• Gymnosperms	> 64	Aves	> 1, 200
• Pteridophytes	>1022	Reptiles	> 399
• Bryophytes	>2584	Amphibia	> 204
• Lichens	>1200	Fishes	> 1, 693
• Fungi	>23000	Insects	>60, 000
• Algae	> 2500	Molluscs	> 5, 000

Biodiversity Hotspots

Internationally recognized Indian biodiversity hot spots are the Himalayan ecosystems in north and the Westernghats in south. There are many internationally recognized aquatic Ramsar sites such as Chilika Lake, Bhittarkanika mangroves in Odisha, rich in aquatic biodiversity with bird fauna. Besides there are many nationally recognized biodiversity hot spots like biosphere reserves, sanctuaries etc.

At International level 18 countries are designated as Mega - biodiversity countries as these countries possess very rich biodiversity (34 hot spots of biodiversity). These countries are: (i) Australia, (ii) Brazil, (iii) Colombia, (iv) Ecuador, (v) India, (vi) Indonesia, (vii) Madagascar, (viii) Malaysia, (ix) Mexico, (x) Peoples Republic of China, (xi) Peru, (xii) Zaire, and the newly added countries are: (xiii) Democratic Republic of Congo, (xiv) Philippines, (xv) Papua New Guinea, (xvi) South Africa, (xvii) United States of America, (xviii) Venezuela. (Figure)

The biodiversity hotspots are extensively distributed in the tropical regions. These regions are located to the vicinity of seas and oceans. Most of the hot spots hold tropical rain forests, Mediterranean-type vegetation, three temperate forests, one tropical Brazil's dry forest, and one semi desert (Succulent Karoo).

About 44% of all species of vascular plants and 35% of all species of the four vertebrate groups namely mammals, birds, reptiles and amphibians are confined to 34 hotspots. It has been found that about 62% of the total surface areas of these hotspots are yet to be protected.

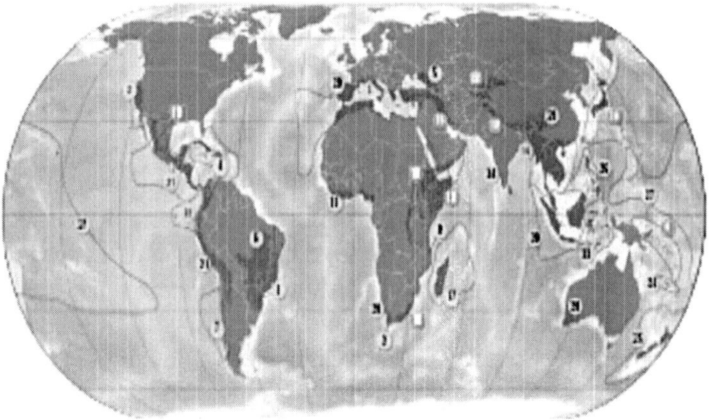

Figure 1 Hot Spots (34) of Biodiversity (taken from WWF)

Huge loss of habitats and fragmentation of habitats in hotspot areas occur. In view of this majority of the endemic species of hotspot areas are considered as threatened species in the International Red lists of species.

The main types of biomes show the diversity of world's ecosystems. Most of these biomes are situated within the 30° N and 30 °S of the Equator.

Biomes

- Different types of ecosystems have evolved by the interaction of climate with parent rock material and the available flora and fauna. Broadly two types of ecosystems such as terrestrial and aquatic have been evolved.
- The ecosystems occurring on a sub continental or on a large land mass basis are referred as Biomes,/Biochores or /Regions.The division lines of these biomes tend to be parallel to the lines of latitude.
- The same type of biome is found within the same general latitudes.This is evident in the case of tundra and boreal forests in both the old and new world.(North Woods- between 45 and 57 north latitudes). Precipitation patterns & wind patterns, solar radiation are latitudinal.
- In mountains, however ,the division lines between biomes is altitrudinal.The biomes found in a given altitude also vary with latitude.
- The zone of Douglas fir- Panderosa pine lies between 675 and 2000 m in Cascade mountains of the northwest but within 1350-2300 m in central and within 1650-2700m in southern Sierra Neveda.
- A particular biome occurs at progressively lower latitudes at progressively northern latitudes.

These biomes are: (i) Tropical forest, (ii) Savanna, (iii) Desert, (iv) High mountain areas, (v) Chaparral, (vi) Temperate grasslands, (vii) Temperate Deciduous forests, (viii) Coniferous forest, (ix) Tundra (arctic and alpine).

Latitudinal & Altitudinal Gradient Influence Biodiversity Distribution: Global pattern of biodiversity is conspicuous on the basis of latitudinal gradient in species richness. The pattern of the average species richness generally increase from high to low latitude, regardless of taxa (namely plants, mammals, birds, reptiles, fishes, insects), geographical position or time domain. Exceptions to this rule may occur, especially with regard to aquatic flora, and parasites and life histories.

Altitudinal -Latitudinal Gradient.

This photograph is taken from different sources

Since seasons, temperature, rainfall, duration of sun light etc vary latitudinally; it is not surprising to observe this biodiversity pattern. Altitudinal gradient also influence biodiversity. Mountain ecosystems show drastic changes in vegetation and biodiversity and climate depending upon the height from the base of the mountain to the summit of mountain. It has been observed that in Norway and the Alps, the species richness of plants decline by 40 for every 100m rise in Elevation. The environmental factors like temperature, precipitation vary with increase in height. Besides the mountain slopes reduce the land area with elevation. In general in low elevation, species diversity is greater than at high elevation, where temperature is cooler and the growing season is shorter. In general, favourable climate, parent rock material and geological stability, water courses and hydrology, land use and geographic isolation make mountain ecosystems in low and medium elevations rich in biodiversity.

Rain forests form the life line of ecosystem balancing. More than fifty percent of world's biodiversity are found in rainforests. Biologists estimate that atleast half of the species occur in tropical rain forests. A tropical rain forest may possess 425-800 kinds of woody trees in one hectare of land in comparison to 5-50 species of woody tree species in middle or temperate region.

The Amazon forest release twenty billion tonnes of moisture every day, helping to water crops around the world. Healthy rainforests absorb about 10% of man's carbon emissions every year. Deforestation releases more CO_2 than all the transport. Saving the rainforests in the world, the Himalayas and the Westernghats in India will be an essential first step. Deforestation is one of the largest contributors to climate change. Rain forests provide countless services to humanity, often unnoticed. (The Prince of Wales- Rainforest Project, 1996). Man has a genius for survival, often only manifest at the eleventh hour. Now is the time when we must act together to avoid the catastrophe of climate change. And arrest biodiversity loss. In view of the discussion made above, it is concluded that *biodiversity forms foundations for* interdependence of organisms. food chain relationships & multichannel food web, maintaining the balance of Nature, maintaining sustained productivity& ecosystem stability, providing food, fodder medicine, shelter, material, fuel, fiber and other utility material & foundation for agriculture, industry and civilization.

Mangroves- wetlands of International Reputation (Ramsar sites) - Bhittarkanika

Wetlands are the interface between land and water systems. They may be natural or artificial areas of marsh or water that could be static or flowing. These wetlands include ponds, lakes, swamps, marshes, mangroves, coral reefs, rice fields, shallow coastal marine waters; sub tidal aquatic beds; rocky marine shore; estuarine waters, intertidal mud, sand and flats; intertidal forested swamps including mangrove swamps, intertidal salt and fresh water marshes; salt pans; Brackish water lagoons.

Inland Wetlands

Rivers & Streams including seasonal rivers; Inland deltas; Riverine flood plains; freshwater lakes, freshwater ponds; Shrub swamps; freshwater swamp forests; Geothermal wetlands.

Manmade Wetlands

Aquaculture ponds; Ponds; Water storage areas-dams-reservoirs, barrages; salt pans, Excavations-gravel pits, mining pools; Irrigated land, rice fields, canals and ditches; Seasonally flooded arable farm land; Wastewater treatment sewage farms, oxidation basins.

Distribution of Indian Wetlands

Himalayan Wetlands- Ladakh Zanskar, Pangong Tso, Morari, Chantau, Noorichan, Chushul and Hanlay marshes.

Kashmir Valley-Dal, Anchar, Wular, Haigam, Malgam, Haukensar and Kranchu Lakes,

Central Himalayas-Nainital; Bhimtal, Naukuchital;

Eastern Himalayas- many wetlands in Sikkim, Assam, Arunachal Pradesh, Meghalaya, Nagaland and Manipur. Beels in Brahmaputra and Barak valley.

Indo-Gangetic Wetlands-Indo-Gangetic flood plains; rivers & rivulets.

Coastal Wetlands- Coastal wetlands in West-Bengal, Odisha, Andhra Pradesh, Tamilnadu, Kerala, Karnataka, Goa, Maharashtra & Gujarat..Important wetlands are Pulicat, Chilika, Kolleru and Vembanad lakes. Mangrove forests of Sundarbans, Bhittarkanika, & of Andaman& Nicobar islands. Offshore Coral reefs in Gulf of Kutch, Gulf of Mannar, Lakshadweep & Andaman & Nicobar Islands.

The Indian village ponds, rivers, rivulets, springs etc. are also wetlands.

Table 4. Wetland Areas of India.

Wetland Areas of the Country : (In million hectares) :			
Rice Cultivation Areas	40.9	Coral Reefs	0.17
Fish culture Areas	03.6	Rivers including main tributaries	28, 000km
Capture Fishery Areas	02.9	Canals & irrigation channels	1, 13, 000km
Mangroves	0.4		
Estuaries	3.9		
Backwaters	3.5		
Impoundments	3.0		
Total Area	58.2		

Sources: Indira Gandhi Institute of Development Research

Ecological Functions of Wetlands:

Water Quality, Hydrology, Flood Control, Carbon Sequestration, Climate Stability, Shore stabilization, Habitat for wildlife, Biodiversity.

Interdependency of Biodiversity & Wetland:

Flowering Plants, Algae, Fungi, Microorganisms, : Mammals 7 Other animals feeding on vegetation, Roosting & resting by birds and animals, plants, Nesting & Breeding by fishes, turtles, amphibia, invertebrates, Different Niches, & Linkage to land-based ecosystems.

Economic Benefits:

Agriculture & Farming, Water Sports, Ecotourism, Water Supply, Timber, thatch, medicine, food, fodder & fuel, Transport, Industrial-Urban Development, Commercial Fishery & Aquaculture, Wildlife Conservation etc.

Human activities threaten Wetlands:

- Alteration of wetlands to other land uses- Agriculture, Urban development, Forestry operation,
- Encroachment for road construction and for railways,
- Increased siltation,
- Pollution from boats, domestic sewage, pesticides, fertilizers and industrial effluents,
- Over Grazing by livestock,
- Overfishing,
- Infestation with aquatic weeds and eutrophication,
- Diversion of water to irrigation, industrial & domestic uses,
- Poaching, illegal trapping and hunting of wildlife,
- Tree cutting for fuel,
- Disturbance from recreational activity and tourism.

India joined Ramsar Convention in 1981 and Convention on Biodiversity (CBD) in 1993 and became a signatory to Natural Heritage Convention.

Table 5. Ramsar sites in India.

Protected under Ramsar Convention:

	Wetland	Date of Declaration	Area
1.	Chilika Lake, (Lagoon), Odisha	1.10.1981	116, 500 ha
2.	Keoladeo Nat. Park, Rajasthan	1.10.1981	2, 900 ha
3.	Wular Lake, J & Kashmir	23.03.1090	18, 900 ha
4.	Loktak Lake, Manipur	23.03.1990	26, 600 ha
5.	Harike Lake, Punjab	23.03.1990	4, 100 ha
6.	Sambhar Lake, Rajasthan	23.03.1090	2, 873 ha
7.	Kanjli, Punjab	22.01.2002	
8.	Ropar, Punjab	22.01.2002	
9.	Sasthamkotta Lake, Kerala	19.08.2002	
10.	Tsomoriri, Jammu & Kashmir	19.08.2002	
11.	Vembanad-Kol Wetland, Kerala	19.08.2002	
12.	Ashtamudi Wetland, Kerala	19.08.2002	
13.	Bhittarkanika, Odisha	19.08.2002	~14, 500 ha
14.	Bhoj Wetland, Madhya Pradesh	19.08.2002	
15.	Deeper Beel, Assam	19.08.2002	
16.	East Calcutta wetlands, West Bengal	19.08.2002	
17.	Kolleru Lake, Andhra Pradesh	19.08.2002	
18.	Point Calimere, Tamil Nadu	19.08.2002	
19.	Pong Dam Lake, Himachal Pradesh	19.08.2002	
20.	Rudrasagar Lake, Tripura	08.11.2005	
21.	Surinsar-Mansar Lakes, Jammu & Kashmir	08.11.2005	
22.	Upper Ganga River (Brijghat to Narora Stretch) Uttar Pradesh	08.11.2005	
23.	Renuka Wetland, Himachal Pradesh	08.11.2005	
24.	Chandertal Wetland, Himachal Pradesh	08.11.2005	
25.	Hokera Wetland, Jammu & Kashmir	08.11.2005	
26.	Nalsarovar Bird Sanctuary, Gujarat		2012 385ha

Bhittarkanika Mangrove Ecosystems

The mangrove forests of Bhitarkanika in the Brahmani-Baitarani-Dhamra River's deltaic region comprise a single continuous and compact forest patch and are the second largest compact patch of mangroves after Sundarbans of West-Bengal.

Being a deltaic region, the area bears large number of creeks and rivulets. The mangrove forests are in different seral stages with 20 different plant associations and the floral diversity includes 62 plant families with 301 species.

The dominant plant families are Poaceae (35 species), Cyperaceae (31 species), Euphorbiaceae (15 species), Pubiaceae (14 species), Papilionaceae (14 species), Rhizophoraceae (10 species.).The common mangrove plant families are listed below.

Mangrove Plant Families of Bhitarkanika

1. Rhizophoraceae, 2. Myrsinaceae, 3. Plumbaginaceae, 4. Sterculiaceae, 5. Meliaceae, 6. Sonneratiaceae, 7. Avicenniaceae, 8. Euphorbiaceae, 9. Aizoaceae, 10. Chenopodiaceae, 11. Hippocrateceae, 12. Tamaricaceae, 13. Malvaceae, 14. Combretaceae, 15. Tiliaceae, 16. Papilionaceae, 17. Caesalpiniaceae, 18. Convolvulaceae, 19. Asciepiadaceae, 20. Apocynaceae, 21. Periplacaceae, 22. Bignoniaceae, 23. Salvadoraceae, 24. Acanthaceae, 25. Verbenaceae.

The Faunal composition of Bhitarkanika

The vertebrate fauna, especially the Reptilian fauna is very rich. They are the King Cobra, Kraits, Python, Salt water crocodile (Crocodiles porous), the Olive Ridley Sea Turtles (Lepidochelys olivacea) etc. The mammalian fauna comprise of Leopard, stripped hyaena, Chital, Deer, Sambar and wild boar.Among the small mammals, the common mongoose, smooth coated Indian Otter, the small Indian Civet, Rey musk Shrew, the stripped pal squirrel, the Indian porcupine, shortnosed fruit bat. The gangetic dolphin, Humpback dolphin, Irrawaddy dolphin and the little Indian Porpoise are found. These aquatic mammals are scheduled animals as per Wildlife (Protection) Act-1972.About 174 species of birds have been recorded from the area. 82 bird species have been recorded to nest & breed here.%7 species of waterfowls have been recorded in the winter season. One of the largest heronries with more than 20, 000 birds have been recorded in this wetland. (Palita, 2012).

Bhitarkanika (145 sq.km area) has been declared as a National Park and since 2002 the park has also been declared as a Ramsar Site. Bordering the site, an area of 1435 sq.km. has been declared as Gahirmatha Marine Wildlife Sanctuary since 1997 and the sanctuary covers the forest blocks of Mahanadi delta mangroves comprising 27 sq.km. Gahirmantha Sea beach, which is the largest rookery for Olive Ridley Sea Turtles in the World.

In Bhitarkanika coast two species of horse shoe crab namely, (1) Tachypleus gigas, (2) Carcinocorpious roticaudata are highly exploited. Considering their medical importance, they may be considered threatened and require conservation measures. This wetland is the major breeding ground for estuarine and brackish water fish; arthropod (prawns), and molluscan fauna and sea turtles along the east coast of India.

Objectives of Wetland Conservation Strategy

1. Conservation & Collaborative Management,
2. Prevention of loss and restoration,
3. Sustainable Use

Action Plan for National Wetland Strategy give importance to (i) Protection, (ii) Planning, monitoring, & management, (iii) Participation of stakeholders, (iv) Inventory preparation and prioritization, (v) Legislation and periodical amendments if required, (vi) Interministerial and cross sectorial coordination and responsibility decentralization, (vii) Capacity building, (viii) International Cooperation, (ix) Public and corporate awareness, (x) Research

Laws to protect Wetlands in India: The following laws are important to protect Indian wetlands (i) The Indian Fisheries Act, 1857, (ii) The Indian Forest Act, 1927, Forest Conservation Act, 1980 amended 1988 and 2006, (iii) Wildlife (Protection) Act, 1972, Wildlife (Protection) Amendment Act, 1991, (iv) Water (Prevention & Control of Pollution) Act, 1974, (v) Territorial Water, Continental Shelf, Exclusive Economic Zone and Other Marine Zones Act, 1976, (vi) Water Cess Act, 1977, (vii) Maritime Zone of India (Regulation and Fishing by Foreign Vessels) Act, 1980, (viii) Environment (Protection) Act, 1986, (ix), Coastal Zone Regulation (Coastal Regulatory Zone) Notification, 1991, (x) National Conservation Strategy & Policy Statement on Environment and Development, 1992.Besides the International Commitments are the following: (a) London Convention (1933) -Convention Relative to the Preservation of Fauna and Flora in their Natural State, (b) Ramsar Convention (1971) -Convention on Wetlands of International Importance Especially as Waterfowl Habitat, (c) Natural Heritage Convention (1972) - Convention concerning the protection of the World Cultural & Natural Heritage, (d) CITES (1973) - Convention of International Trade in

Endangered Species of Wild Fauna & Flora, (e) MARPOL (1973/1978) -International Convention for the Prevention of Pollution from Ships, (f) Bonn Convention (1979) -The Convention on the Conservation of Migratory Species of Wild Animals, (g) UNCLOS (1982) United Nations Convention on the Law of the Sea, (h) Convention on Biological Diversity (CBD) (1992).

Rural Livelihood based biodiversity

In developing countries the livelihood options and rural economy are largely bio-resource based, especially on: Agriculture & horticulture, (crops & fruits), Forestry (firewood, minor forest products), Home gardens (vegetables, fruits, flower), Diary (milk, cow dung, ploughing), Fishery (fish, prawn, turtles), Goatery (meat, milk, skin), Poultry (meat and egg), Artisan work based on bamboo, wood, cane, grass and other resources. Traditional ecological knowledge and bio- resource management is important for rural systems. Biodiversity provides us food, shelter material, raw material for medicines, gases for survival, other ecosystem services like recycling of water and other materials

We need natural protectors around us, like the mangroves of Bhittarkanika (Odisha), & Sunderban (West Bengal) as coast guard and the "green-friends" (jungles) like Similipal, Himalayas, Westernghats, and Easternghats, and Rain forests to generate and supply life supporting oxygen or sequester the atmospheric carbon preventing global warming.

Agricultural Biodiversity

In addition to biodiversity in the wild, the biodiversity of organisms used for farming and other human activities is very important. In agriculture, 7 000 species of plants are used by farmers in the world, but only 30 species provide 90 percent of our calorific intake. The top three crops are wheat, rice and maize (corn) with around 500 million tons annual production each. Data from various sources indicate that India has contributed about 167 cultivated plant species (7%) to the world. Within these dominant crop species, there are many hundred thousands of varieties (landraces, cultivars) adapted to local climates, farming practices, and cultural predilections like taste, color, structure, ability to store the products etc. Much of this large crop diversity is important for providing the initial material for breeding.

Biotechnology and Biodiversity

World's population is about 7 billion (7, 000 million) at present and has been projected to grow to about 9 billion by 2050. Poverty is a major concern in the world. Assuming perfect distribution it is projected that food production at current level would feed about 7 billion people. It is crucial that food production keeps pace with the increasing population.

To feed a huge population, the food production has to be doubled. This will be a tall order as water is scarce, additional agriculture land may not be available and constraints in other factors including the climate change. For the improvement of food security and alleviation of poverty, the proponents of genetically modified crop believe that the use of this technology would help reduce poverty and improve food security. The only answer at present is to adopt new farming systems and modern technology to increase food production. Availability of sufficient food and nutrition, and bio-energy and environmental protection are now universal concerns. In this context, use of biotechnology to increase crop productivity and improve crop quality from nutritional angle may be the only answer. The genetic modifications of cereals, pulses, fruits and vegetables are required not only to increase productivity, but also for better nutrition. Biotechnology is also a tool to increase biodiversity and to conserve biodiversity. The bio-safety aspects of genetically modified food are an emerging field for research and require immediate attention.

Scientists and farmers have been changing the genetic structure of plants and animals for thousands of years to increase food productivity and to get desired animals for human use for various purposes. Vertical gene transfer between cultivars and wild plants has always occurred within the limits of species, if the two types of plants were in close proximity and flowered at the same time. Hybrid production is age old practice. When apple cultivators take a an apple tree that produces juicy and sweet apples and crossbreed it with an apple tree that is resistant to frost or a particular disease, the cultivators produce a hybrid, which is a genetically modified apple tree to produce juicy and sweet apples even under the adverse condition of frost. This manipulation happens in nature and man acts as an agent, like crosspollination by bees or insects or the wind.

But modern biotechnology deals with altering genetic material, the DNA in an organism in a way that does not occur naturally. The genetically changed organism is called genetically modified organism or GMO. Genetically medication permits desirable genes to be transferred to another plant to create a GMO- crop that can be productive, more nutritious, and have resistance to herbicides and pests. The following examples will explain the situation.

Example 1: In Europe strawberries was genetically modified to withstand severe cold conditions that prevail in Europe. Biotechnologists isolated an anti-freezing protein, found in the DNA of a fish called the winter flounder and inserted into the DNA of strawberries. This altered DNA was then passed on to subsequent generations of strawberry plants. The purpose was to get the benefits of the introduced gene without taking away the good attributes of strawberries. Thus the GM-strawberries should withstand the severe cold condition but does not taste fishy. Ethical questions are involved. But from scientific angle and from the angle of nutritious food production, this appears to be an excellent solution if the GMO-food does not create health problems.

The World Heath Organization estimates that the world's population will exceed nine billion by 2050. There will be severe scarcity of food unless we increase food production. In view of this biotechnology is the answer.

Example 2: In United States of America, a GM-crop such as golden rice, that has been trialed by combining rice with beta-carotene. Beta-carotene intake prevents blindness in children by supplying vitamin-A.

Another example in production of Bt-corn, Bt-maize or Bt-cotton, which involves the transfer into the crops of a gene of a bacterium (*Bacillus thuringiensis*) that naturally releases an insecticidal toxin which gives protection to crops from bollworms that eat cotton leaves and many other crop plants. GM crop technology can improve yield of important staple crops by creating varieties of crops which are disease resistant and which can withstand extreme climatic conditions such as low rainfall and uneven soil fertility. In Mexico, for example through genetic engineering, it was possible to create GM maize that was able to grow in acidic soils. This has helped the poor Mexican farmers who were able to grow GM maize that produced 30% more yield and thus improved their economic condition by increasing profits.

Other notable examples are virus resistant GM potatoes in Colombia and Kenya, insertion of genes against aluminum toxicity in Mexican wheat.

Research is being conducted to produce GM crops that could produce bio-pharmaceuticals. Examples of such crops are potatoes that have the cholera causing bacterial gene inserted to produce tiny amounts of cholera toxin that immunize against cholera or tomatoes engineered to carry plant-derived vaccines. The advantage of edible crops with vaccines is that they would be cheaper to grow locally. Biotechnology tools are now used to conserve biodiversity and also to create new biodiversity.

Soil Biodiversity

Soil organisms include microorganisms, such as bacteria, actinomycetes, fungi and algae, and soil animals, such as protozoa, helminthes worms, nematodes, annelid worms (enchytraeids and earthworms), collembola, mites, and insects and their larval forms, pupae, arachnids, other arthropoda, gastropods, etc. Some mammals and birds use the soil for various purposes. They play an important role in determining the physical and chemical properties of soil, soil fertility, and so on. Earthworms and termites play a significant role in soil formation. Earthworms may burrow two metres into the soil, make numerous transverse burrows and produce worm casts by eating the soil and soil organic matter. These casts are very rich in nitrogen and mineral substances, and increase soil fertility. The burrows facilitate aeration and increase water-holding capacity. Dash and Patra (1979) measured 77 tonnes of dry weight of earthworm cast production per hectare per year in an Indian grassland site. Dash et al (2010), Dash and Dash (2009), and Edwards and Lofty (1972) have estimated that earthworm cast production ranges from 2 to 247 tonnes per hectare per year in different world sites. This is equivalent to bringing up layers of soils between 1 to 5 cm thick to the surface every year. In addition, a large amount of soil is deposited in burrows as casts. Thus the soil turnover provides a stone-free layer about 15 cm deep on the surface. Earthworm casts contain more water-soluble aggregates than the surrounding soils.

The following three tables show data on the soil faunal diversity in India and the role play by soil biota in the soil processes.

Soil Fauna Diversity in India

Taxa	Total No. of species		Endemic species	Geographical distribution
	(World)	(India)		
Echiura	127	43	12	Coastal areas
Chilopoda	3000	156	NA	South India, N.E India, Mumbai, Gujarat, Bihar, UP, HP & Orissa.
Diplopoda	7500	162	NA	
Orthoptera	17250	1750	200	Almost entire India
Dermoptera	2000	320	-	Almost all over India
Embioptera	200	33	14	Discontinuous distribution: Himalayas, Gangetic plains, Barkuda Island, Eastern Ghats, Western Ghats).
Isoptera	2000	253	170	Almost entire India
Hemiptera	NA	6500	1335	Restricted distribution: Eastern Himalayas, Western Himalayas, Peninsular India.
Neuroptera	5000	335	262	Himalaya, Desert, Peninsular India, Insular ecosystem
Coleoptera	8,00,000	15500	3113	Widespread distribution
Mecoptera	350	15	-	Eastern India, UP, Maharashtra
Diptera	100,000	6093	2132	Widespread
Trichoptera	7000	812	649	Himalayan and Peninsular India.
Annelida	12000	1000	-	Widespread

Source: ZSI, Kolkata

Relationships between soil invertebrate behaviour and soil physical process

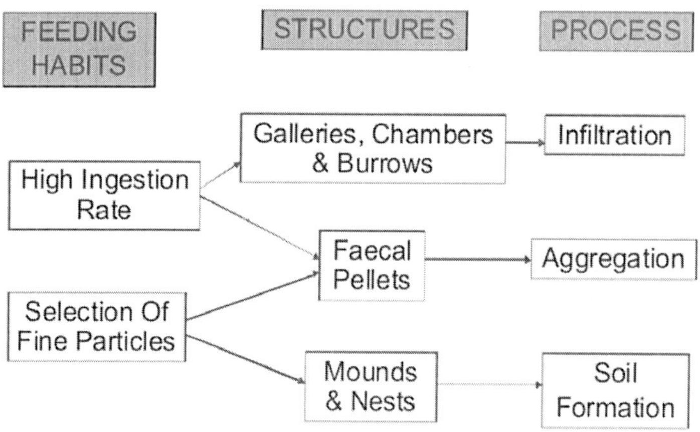

Influences of Soil Biota on Soil Processes in the Ecosystem

	Nutrient Cycling	Soil Structure
Microflora	Catabolize organic matter, Mineralize and immobilize nutrients in aggregates	Produce organic compounds that binds agreements, Hyphae entangle particles unto aggregates
Microfauna	Regulate bacterial and fungal populations after nutrient turnover	May affect aggregate structure through interactions with microflora
Mesofauna	Regulate fungal and microfaunal populations, Alter nutrient turnover, Fragment plant residues	Produce fecal pellets, create biopores, promote humification
Macrofauna	Fragment plant residues, Stimulate microbial activity, Redistribute organic matter and micro-organisms	Mix organic and mineral particles, Create biopores, promote humification, produce fecal pellets

Microbes form a vibrant living community in the soil contributing to a number of nutrient transformations and ecosystem functioning. They are involved in organic matter decomposition, N_2-fixation, solubilization and immobilization of several major and minor nutrients (Alexander, 1971). Microorganisms play important role in soil structure maintenance, plant growth promotion through secretion of hormones and soil borne disease control. The diversity and richness of soil microorganisms are huge and subject of interests to scientists over the years. But till today much is yet to be is known on the complex living biota in the soil and their biophysical, biochemical functions and interactions in the soil ecosystem. But during the last 50 years, many beneficial effects of microbes in temperate and tropical soils have been discovered. We have been making use of microbes for improving productivity in agriculture, industry and pharmaceuticals and they are used in biotechnology.

The following diagram shows the Indian situation with regard to endemic and exotic species

IUCN Red Data Book

The International Union of Conservation of Nature has categorized different species depending upon their occurrence in nature.

The categories are listed below.

EXTINCT (X) : No longer existing in the wild.

ENDANGERED (E) : In immediate danger of extinction if factors causing decline continue operating.

VULNERABLE (V) : Likely to become endangered soon if factors causing decline continue operating.

CRITICALLY RARE (R) : Not endangered or vulnerable at present, but might be at risk by localized threats.

INDETERMINATE (I) : A temporary grouping when not enough information is available to put into endangered, vulnerable or critically rare categories, but it is reasonably certain that the species does indeed belong in one of these categories. Further data are needed to determine the status.

UNCERTAIN (U) : A temporary grouping for unconfirmed reports suggesting inclusion in endangered, vulnerable or critically rare categories.

NO LONGER THREATENED: No longer included in the above categories due to an increase in population sizes or numbers, or due to the discovery of more populations or more individuals.

The Protea Atlas Project is currently updating the Red Data List status for all our protease. If you want to help, please contact us.

The above categorizations are not precisely defined, and a new categorization has been made by the IUCN. The aim is to be uniform, objective, and transparent. Consequently the new definitions are long and involved, but in future, all categorizations will specifically state their reasoning (in summary as A2 or B3 based on the criteria involved).

Thc Ncw Categories are (these are condensed summaries - please see the IUCN web site for fuller details) :

EXTINCT (EX) : A taxon is Extinct when there is no reasonable doubt that the last individual has died.

CRITICALLY ENDANGERED (CR) EXTINCT IN THE WILD (EW) survives in cultivation, in captivity or as a naturalized population (or populations) well outside the past range. A taxon is presumed extinct in the wild if exhaustive surveys in its known habitat in its historical range have failed to record an individual.

A taxon is Critically Endangered when it is facing an extremely high risk of extinction in the wild in the immediate future.

A. Range reduction of 80% over 3 generations or 10 years;

B. Occurrence of less than 100sq km or occupancy of less than 10sq km and decline in populations;

C. Less than 250 plants and decline;

D. Less than 50 mature plants;

E. Probability of extinction of 50% in 10 years or 3 generations.

ENDANGERED (EN) A taxon is endangered when it is not Critically Endangered but is facing a very high risk of extinction in the wild in the near future.

A. Range reduction of 50% over 3 generations or 10 years;

B. Occurrence of less than 5000 sq km or occupancy of less than 500sq km and decline in populations;

C. Less than 2 500 plants and decline;

D. Less than 250 mature plants;

E. Probability of extinction of 20% in 20 years or 5 generations.

VULNERABLE (VU) A taxon is Vulnerable when it is not Critically Endangered or Endangered but is facing a high risk of extinction in the wild in the medium-term future.

A. Range reduction of 20% over 3 generations or 10 years;

B. Occurrence of less than 20 000 sq km or occupancy of less than 2 000 sq km and decline in populations;

C. Less than 10 000 plants and decline;

D. Less than 1 000 mature plants, or low occupancy (<100km2), or less than 5 locations;

E. Probability of extinction of 10% in 100 years.

LOWER RISK (LR) A taxon is Lower Risk when it has been evaluated, does not fall into the categories Critically Endangered, Endangered or Vulnerable. There are three subcategories

Conservation Dependent (cd). Taxa which are the focus of a continuing conservation programme, and which would become threatened if such actions stopped.

Near Threatened (nt). Taxa which do not qualify for Conservation Dependent, but which are close to qualifying for Vulnerable.

Least Concern (lc). Taxa which do not qualify for Conservation Dependent or Near Threatened.

DATA DEFICIENT (DD) A taxon is Data Deficient when there is inadequate information to make an assessment based on its distribution and population status.

NOT EVALUATED (NE) A taxon is Not Evaluated when it is has not yet been assessed against the criteria.

If you want more information about the new categories please visit.

Conservation of Biodiversity

Conservation is being done either by *in situ (*either in the natural or semi-natural habitat) or *ex situ* (in some purpose-built environment). The choice is contextual.

In situ conservation involves maintenance and protection of natural habitats.

Ex situ conservation involves botanical gardens; zoo and seed banks/germ-plasm bank. Both processes require precise knowledge of taxonomy.

This would require a selection of the most appropriate areas, called the hot spots. Conservation also embraces various components of agro-biodiversity like crop varieties, land races, semi-domesticates and crop relatives. The role of indigenous communities in maintaining agro-biodiversity is stressed by the Global Biodiversity Assessment and the Leipzig Plan of Action, two recently concluded international agreements. United Nations convened the CBD (Convention of Biological Diversity) and succeeded in having it adopted in 1992. It entered into force in 1993. CBD requires that all Member States take measures to conserve both native and agricultural biodiversity.

As the human population expands, the need for food is expected to double in the next 30 years. Increasing crop productivity on the land already under cultivation would prevent or at least reduce habitat destruction.CBD recognizes that States have a sovereign right over biodiversity within their own territory. Living organisms or their products may, under the terms of the CBD, only be removed from a country under mutually agreed conditions. The CBD is a comprehensive approach to biodiversity conservation of both wild and domesticated species.

Scientific data shows that the biological diversity has increased slowly but steadily through the evolutionary process over geological time. However, mass global extinction of species has occurred due to natural processes of glaciations, volcanic eruption, earthquake etc.At present we loose on the average per year one species of mammal, one bird species and many invertebrate species from the planet. This realization has helped to develop the concept of environmental management and planning so that in the scenario of accelerating economic development, the biodiversity and quality environment is maintained and a sustainable society is evolved.

Remedial measures

It is important to identify unique ecological pattern and conserving them and to create biotic resource inventory. Periodic monitoring (Biological, Chemical and Physical, remote sensing etc.) is very important to know the status. Identifying endangered, rare, vulnerable, and threatened species and conserving them is another important aspect. Biotic salvage operations include restoring wilderness, habitat and niches, restoring food webs, understanding the home range and territorial requirement, doing captive breeding and understanding their fitness values and establishing Biospheres, National Parks, Sanctuary, Zoological Gardens etc. and Gene banks. Creating awareness by environmental education (curricula, media, folk dance, jatra (opera) etc.).

Environmental laws are a good instrument to protect biodiversity. Law implementation is important. 'No Development" without adequate environment protection and addressing human issues should be allowed.

Action Plan: The Biodiversity Act

The Biological diversity Act, 2002 received assent of the President of India on the 5th February 2003. It is expected that this Act will be a valuable instrument for Biodiversity conservation and to achieve sustainable development. Traditional knowledge is also used as an input into Pharmaceuticals, Ayurved medicines, Agriculture, Biological pesticides, cosmetics and toiletries etc.The traditional knowledge is being widely disseminated and commercially exploited with only a small proportion of the benefits flowing back to the people/ communities.

Questions arise who owns traditional knowledge patenting? It is expected that the Biological Diversity Act has considered this aspect also.

The India Seed Act and Patent Act should be reviewed keeping Biodiversity Act in view and the interests of Indian farmers, Industry and people's interest.

India has taken important steps like : (i) Setting biotic resource inventory, (ii) Identifying the endangered, vulnerable, rare and threatened species and ecosystems (iii) Identifying disease causing organisms, vectors etc. and devising methods for their eradication or control in the context of environmental management, (iv) Biotic salvage operations for ecosystems and threatened species populations, (v) Creating trained man power for taking up ecosystem analysis especially from the biological damage control point of view and for sustained yield point of view. The problems however, are extremely complex.

We should be compassionate for biodiversity conservation for the sake of continuance of man on this planet? World Trade Human rights, Traditional knowledge on alternate medicine and other aspects are linked to biodiversity.

Ecological, genetic, social, economic, scientific, educational, cultural, recreational and aesthetic values attached to biodiversity. Many types of life forms with multi channel food webs help in decreasing system perturbation.

Data from various sources indicate that the total cost of benefits from a hectare of tropical deciduous forest with regard to oxygen production, prevention of soil erosion, wildlife conservation, providing shelter & food to birds & other animals, controlling air pollution and playing role in water & mineral cycling may account for atleast three to five lakh rupees per year with forest remaining intact. Biodiversity is renewable as it has a life cycle and continuous harvest is possible provided effective and efficient management care is taken to maintain them. Management of biodiversity and resources is crucial for sustained production of materials for a sustainable society. In view of this resource conservation/ damage rehabilitation cleaner production technology, adoption of recycle and reuse and ecofriendly waste disposal methods, Environmental Impact Assessment/Environmental Management Programme aspects under Environment Protection Act-1986 are to be given importance on priority basis.

Livestock as Renewable Resources: Livestock are important resources, as they provide milk, eggs, meat, skin (leather) and other products. The total livestock in India mount to a huge population consisting of more than 185 million cattle, 118 million goats, 61 million buffaloes, 45 million sheep, 1 million horses and ponies, 1 million camels and about 1 million other livestock. Besides, there are 156 million tonnes poultry, fowls, ducks, turkeys, etc. They provide about more than 1 million tonnes of meat, 74 million tonnes of milk, 40 million kg of wool and 13 million eggs. The amount of dung produced is huge and is used for biogas production and as manure. Livestock are also used for rural transport and agriculture. To get a sustained yield, their proper breeding, health care and management and diet are essential. Recent advances in biology have helped produce better livestock. The milch breed (milk producing) cattle are the Gir, Red Sindhi, Deoni and Sahiwal and the draught breed (work animals) cattle are the Nagari, Malvi, Kenkatha, Khillari, Hallikar, Amrit Mahall, Ponwar, and Siria.

The total milk production has increased from 17 million tonnes to about 200 million tonnes . Some examples of dual-purpose (both for milk and work) cattle are the Rewati, Kankrej, Ongol and Danji. The humped zebu cattle originated in India and there is evidence that humpless cattle were domesticated in India during the Harappan period.

India has the best water buffaloes in the world. These buffaloes are the Murrah, Surti, Nili-Ravi, Jaffarabadi and Mehsana. The common breeds of Indian sheep are Kashmir, Chokla, Magra, Kali, Gaddi and Bhakarwal. Breeds of Merino and Suffolk sheeps have been imported for production of better wool.

Milk breeds of goat, such as Saanan, Alpore, Nubian and Toggenberg, have been introduced to develop better breeds to suit the Indian environment. The wool and textile breeds are Ashman and Chegu, Beetal, Jamunapari, Barbati, Deccan, Black Bengal and Malbari. The niche requirements of sheep and goats are different. Livestock contributes at least fifty thousand million rupees a year to our national economy. Of this, poultry contributes more than ten thousand million rupees. Almost all the world's fowls trace their origin to India's red jungle fowl (Gallus gallus). Ducks constitute about 10% of the total poultry in India. In Indian, we have the Assel, Kadaknath, and short-legged Nicobari breed fowls. The important broiler strains are B 77, JBL 80, IBB 83, ILI 80 and ILI 82.

Aquaculture as a Renewable Resource: Fish are an excellent source of protein. They feed on plankton (phytoplankton and zooplankton), weeds and organic matter. Crabs, prawns, mollusks, etc. are also aquatic organisms and are a very good source of animal protein for human beings. Over half the population in the developing countries gets at least 40% of its animal protein food from fish (FAO, Status of Food and Agriculture, 1984, Rome). The natural pearl industry depends upon oyster culture. Fishery includes capture fishery and culture fishery. Capture fishery is usually practiced in oceans, large lakes and other aquatic ecosystems. Culture fishery is usually practiced in inland water bodies and artificial ponds.

The global marine fish catch is on the increase. India's marine fish catch was about 1.2 million tonnes in 1984 and it has increased to 4.1 million tonnes in 2014. India has a very long coastline (about 5, 600 km) and a continental shelf of about 2, 59, 000 km^2. Thus there is a vast area available for tapping. Marine resources include the Bay of Bengal, Arabian Sea, and Indian Ocean, numerous gulfs, coral reefs and mangrove swamps. The estuaries of major rivers, backwaters, lagoons, and brackish water lakes like Chilka, add to India's marine resources. They form an additional 2.6 million hectares. There is a good possibility of increasing the fish catch along the Indian coast using the sophisticated scientific techniques now available. In 1984, India produced about 3 million tonnes of fishery products, occupying the eighth position in the world. The fish production has increased considerably to about 4.2 million tonnes and it has been estimated that this production can be increased to about 12 million tonnes per year (the estimate of India' production in 2015-2016 is > 10 million tonnes, both marine capture and fresh water culture and capture). However, fishing remains confined to a narrow coastal region of 16 km and the rich offshore and deep waters remain poorly exploited. India has about 1.6 million hectares of inland waters. It has many major river systems, such as the Ganges, Jamuna, Brahmaputra, Narmada, Mahanadi, Cauvery and Krishna. Besides, there are canals, ponds, lakes and irrigation channels where culture fishery can be practiced. The main fresh water fish are carps (*Labeo rohita*, other *Labeo sp.*, *Catla catla, Carbine mrigala*, etc.). The exotic carps available are the Chinese carp (*Cyprinus carpio*), green carp (*Ctenophryngodon idella*), mirror carp (*Hypophth-almichthys molitrix*), catfish, and so on. Some Clupeids like *Hilsa* are also found in river systems like Hoogly, Subarnarekha,

Mahanadi, and Chilka Lake. The total fresh water fish production is about 1.77 million tonnes (1996), (about 50% of the total catch). Now culture fishery has been started in pond cultures, artificial enclosures and net pens, providing fertilizers such as cow dung, domestic and agricultural waste and other animal excreta. Pond culture can be: (i) Monoculture, involving one species only; usually trouts, eels and cat fishes, (ii).Polyculture, involving the culture of more than one aquatic species, common combinations are milk, fish with prawn, Chinese carp with Indian carp, and carps with *Tilapia*. Besides, intensive rearing pond culture with high stocking density and water quality control and rich artificial food, eels, trouts, and catfish are usually grown in such cultures. An extensive rearing pond where stocking density is low and food is natural is also practiced. In semi-intensive rearing ponds where natural food supplemented with artificial food is given; usually fertilizers are used to increase food production. Carps, *tilapia*, trout, salmon and some other fish are also cultured in net pens. Milkfish, mullets and some other fish are cultured in enclosures or bamboo fences. In India fish culture is also done in irrigation canals, paddy fields and other small wetlands. Air breathing fishes, such as *Channa* sp., *Clarias* sp., *Heteropneustis*, *Anabas* sp., and so on are cultured in swamps and cage culture systems. A study of the food chain relationships of fishes with a view to shortening the chain is important for aquaculture. Recently, technologies have been developed in Odisha to culture prawns (*Paeneus monodon* and *Paeneus indicus*) in confined salt water tanks. The average aquaculture production is about 600 kg per hectare per year in India. This yield can be increased to 11 tonnes per hectare per year using composite fish culture (using surface, column, and bottom feeding fish in the same pond and thus avoiding competition among them for food). The yield of air breathing fish in swamps is now 15 tonnes per hectare per year.

References

Dash, M.C. and S.P. Dash.2009. Fundamentals of Ecology, McGraw-Hill Publ., New Delhi.

Besides many papers and books were consulted to write this article.

4

History of Sacred Groves and their Role in Maintaining Socio-Religious Unity and Ecosystem Conservation with Examples from Odisha, India

Introduction

Sacred groves comprise of patches of forests or natural vegetation consisting of a few trees to larger areas of forests– that are usually dedicated to local folk deities or tree spirits (Vana Devatas) or a particular god or goddess in Indian traditions.

These areas are protected by local communities because of their religious beliefs and traditional rituals (Figure 1) that run through several generations. This is the socio-cultural concept of the sacred groves. The local community visits the site many times in a year to offer puja to the deity, and for get-together of the community. Most of the sacred groves are either located near a natural water source or have man-made fresh water ponds and dug wells. The degree of sanctity of the sacred groves varies from one grove to another.

In some groves even the dry foliage and fallen fruits are not collected. People believe that any kind of disturbance may offend the local deity, causing diseases, natural calamities or failure of crops. The sacred groves have important conservation values. For example, the *Garo* and the *Khasi* tribes of northeastern India completely prohibit any human interference in the sacred groves. In other sacred groves, deadwood or dried leaves may be picked up, but the live tree or its branches are never cut. For example, the *Gonds* of central India prohibit the cutting of a tree but allow fallen parts such as twigs, foliage and fruits to be used by human beings. The socio-cultural and environmental protection values of sacred groves are immense. This concept keeps people united in all occasions to protect the site and thus the environment is protected. The concept is prevalent now in ancient communities throughout India. In brief a sacred grove is any grove of trees of special socio-religious importance to a particular culture/community inculcating socio-religious unity and furthering the notion of environmental protection.

Figure 1. Kalpeswar Sacred Grove in Sambalpur district of Odisha.People tie a red cloth, offer fruits, milk products to the deity (Shiva Lingam) and ring the bell wishing Lord Shiva's blessings for specific work. People protect the grove, and the water source, which recharges the ground water.

2. History of Sacred Groves in the world

Sacred groves were most prominent in the ancient Near East, and in India (Temple groves/Rishi Ashrams in ancient India of ancient sages (such as Vasistha, Balmiki and others) but feature in various cultures throughout the world. Sacred groves were important features of the European mythological landscape and cult practice of Greek, Baltic, Celtic, Germanic, Near Eastern, Roman, and Slavic polytheism. They were also used in far-east i.e. Thailand, Malaysia, and Japan, and also in some parts of Africa.

2.1 Sacred Groves in India

In India, sacred groves are located all over the country since the dawn of Indian civilization. There is mention of Rishi Ashrams with forest patches, water source, wild animals, and place of worship in scriptures.However, sacred groves do not enjoy protection via Central or Provincial legislation. People and some NGOs work with local villagers to protect such groves. Each grove is associated with a presiding deity, and the groves are referred to by different names in different parts of India. They are maintained by local communities with hunting and logging strictly prohibited within the sacred grove. While most of these sacred deities are associated with local Hindu gods, sacred groves of Islamic and Buddhist origins are also known. Sacred groves occur in a variety of places - from scrub forests in the 'Thar Desert' of Rajasthan maintained by the Bishnois, to rain forests in the Kerala Western Ghats. Himachal Pradesh in the North and Kerala in the South of India are specifically known for their large numbers of sacred groves. The Kodavas of Karnataka maintained over 1000 sacred groves in Kodagu alone. (Down to Earth publications).

Around 14, 000 sacred groves have been reported from all over India (Table 1), which act as reservoirs of rare fauna, and more often rare flora, amid rural and even urban settings. Experts believe that the total number of sacred groves in India could be as high as 100, 000. Ritualistic dances and dramatizations based on the local deities that protect the groves are called *Theyyam* in Kerala and *Nagmandalam*, among other names in Karnataka. There are sacred groves in Ernakulam region in a place named Mangatoor in Kerala. In Central India and Odisha, tribal communities have developed many sacred groves and protect them.

Threats to the groves include urbanization, over-exploitation of resources, and environmental destruction from religious practices. While many of the groves are looked upon as abode of gods, but in the recent past a number of them have been partially cleared for construction of shrines and temples (Guha, 2000, Malhotra et al, 2001).

2.2 Sacred Groves in Europe

Examples of sacred groves in Europe include the Greco-Roman *temenos*, the Norse *hörgr*, and the Celtic *nemeton*, which was largely but not exclusively associated with Druidic practice (Ronald Hutton, *The Druids*, p97 ISBN 978-1-85285-533-8)

There was a common practice of building churches on the sites of sacred groves. English folklore and fairytales mention about the sacred groves and ancient holy trees still exist in the countryside of England. There are two mentions on this tradition in the Bible: Abraham planted a grove in Beersheba, and called there the name of God.—Genesis 21:33 (.http:// en.wikipedia.org/wiki/sacred_groves_of-India). The most famous sacred groves in mainland Greece was the oak grove at Dodona. Outside the walls of Athens, the site of the Academy was a sacred grove of olive trees, still recalled in the phrase "the groves of Academe. (Herodotus, v.119") and in central Italy, the town of Nemi recalls the Latin *nemus Aricinum*, or "grove of Ariccia", a small town a quarter of the way around the lake (Wikimedia Commons has media related to *sacred groves*). In Antiquity, the area had no town, but the grove was the site of one of the most famous of Roman cults and temples: that of Diana Nemorensis, (*The Golden Bough*- James Frazer, 1922). A sacred grove behind the House of the Vestal Virgins on the edge of the Roman Forum lingered until its last vestiges were burnt in the Great Fire of Rome in 64 CE. In the town of Spoleto, Umbria, two stones from the late third century BC, inscribed in archaic Latin, established punishments for the profanation of the woods dedicated to Jupiter (*Lex Luci Spoletina*) have survived; they are preserved in the National Archeological Museum of Spoleto. (National Archeological Museum of Spoleto website).The *Bosco Sacro* (literally *sacred grove*) in the garden of Bomarzo, Italy, lends its associations to the uncanny atmosphere. The city of Marseilles, a Greek colony, had a sacred grove so close by it that Julius Caesar had it cut down to facilitate his siege. In *Pharsalia*, the poet Lucan dramatized it as a 'place where sunlight could not reach through the branches (canopy), where no animal or bird lived, where the wind did not blow, but branches moved on their own, where human sacrifice was practiced, in a clear attempt to dramatize the situation and distract from the sacrilege entailed in its destruction'.

Sacred groves feature prominently in Scandinavia. The most famous sacred grove of Northern Europe was at the Temple at Uppsala in Old Uppsala, where every tree was considered sacred - described by Adam of Bremen. The practice of blót - the sacrificial ritual in Norse paganism was usually held in *lund*s or sacred groves. According to Adam of Bremen, in Scandinavia, pagan kings sacrificed nine males of each species at the sacred groves every ninth year. (http://scandinavian.wisc.edu/mellor/myth/pdf_files/AdamBremen.pdf)

The Celts used sacred groves, called *nemeton* in Gaulish, for performing rituals for the worship of the deity Nemetona - a Celtic goddess. Existence of such groves have been found in Germany, Switzerland, Czech Republic and Hungary in Central Europe, in many sites of ancient Gaul in France, as well as England and Northern Ireland. Sacred groves had been plentiful up until the 1st century BC, when the Romans attacked and conquered Gaul. One of the most well known *nemeton (*meaning a quadrangular space surrounded by a ditch enclosed by wooden palisades) sites is that in the *Nevet* forest near Locronan in Brittany, France. Gournay-sur-Aronde (Gournay-on-Aronde), a village in the Oise department of France, also houses the remains of a *nemeton*. (Venceslas Kruta, et al, 2000) and Maurice Meuleau, 2004)

2.3 Sacred Groves in United States of America

In United States of America, the Bohemian Grove, located at Bohemian Avenue, in Monte Rio, California, is a sacred grove belonging to a private San Francisco-based Men's Bohemian Club. In mid-July each year, Bohemian Grove hosts a two-week, encampment of some of the most powerful men in the world, where they perform symbolic rituals, such as Cremation of Care.

2.4 Sacred Groves in Africa

In **Nigeria**, the concept of sacred groves is present in mythology as well. The Osun-Osogbo Sacred Grove, containing dense forests, is located just outside of the city of Osogbo, and is considered as one of the last virgin high forests in Nigeria. It is dedicated to the fertility goddess in Yoruba mythology, and is dotted with shrines and sculptures. The grove was declared a UNESCO World Heritage Site in 2005. Oloye Suzanne Wenger, an Austrian artist, helped revive the grove.

In Ghana, sacred groves are also present. One of Ghana's most famous sacred groves - the Buoyem Sacred Grove - and numerous other sacred groves are present in the Techiman Municipal District and nearby districts of the Brong Ahafo Region. They provide a refuge for wildlife which has been exterminated in nearby areas, and one grove most notably houses 20, 000 fruit bats in underground caves (UNO). The capital of the historical Ghana Empire, contained a sacred grove called *al-gâba* (Ar. "the forest") for performing religious rites of the Soninke people. Other sacred groves in Ghana include sacred groves along the coastal savannahs of Ghana (Michael O'Neal Campbell, 2004). Many sacred groves in Ghana are now under federal protection - like the *Anweam Sacred Grove* in the Esukawkaw Forest Reserve (Boakye Amoako-Atta-UNESCO). Other well-known sacred groves in Ghana include the *Malshegu Sacred Grove* in Northern Ghana - one of the last remaining closed canopy forests in the savannah regions, and the Jachie sacred grove (Dorm-Adzobu, et al, 1991).

2.5 Sacred Groves in Far East

Sacred groves are located all over Japan and usually associated with shrines. They have existed since ancient times and shrines are often built in the midst of preexisting groves. The Cryptomeria tree is considered sacred and venerated in Shinto religious practice. Among the sacred groves associated with such *jinja*s or Shinto shrines is the 20-hectare wooded area associated with Atsuta Shrine (Atsuta-*jingû*) (Kruta, 2000,*)* at Atsuta-ku, Nagoya. The 1500-hectare forest associated with Kashima Shrine was declared a "protected area" in 1953 (CBC technical Series No. 36).At present, it is part of the Kashima Wildlife Preservation Area. The woods include over 800 kinds of trees and varied animal and plant life cited in 'A guide to Japan's sacred forests (kateigaho.com.). Tadasu no Mori is a general term for a wooded area associated with the Kamo Shrine, which is a Shinto sanctuary of 12.4 hectare near the banks of the Kamo River in northeast Kyoto. (Terry, Philip, 1914). This site is preserved as a national historical site.

(Shimogamo-jinja) The Kamigamo Shrine and the Shimogamo Shrine, along with other Historic Monuments of Ancient Kyoto (Kyoto, Uji and Otsu Cities), have been designated World Heritage Sites since 1994.

The Utaki sacred groves is associated with burial grounds) on Okinawa and associated with Ryukyuan religion, and also associated with toun or *kami-asagi* - regions dedicated to the gods where people are forbidden to go. Sacred groves are often present in such places, as also in Gusukus - fortified areas which contain sacred sites within them (Asato Susumu) The Seifa-utaki was designated in 2003 as a UNESCO World Heritage Site. (Wonder Okinawa: World Heritage site inclusion, 2003). It consists of a triangular cavern formed by gigantic rocks, and contains a sacred grove with rare, indigenous trees like the Kubanoki (a kind of palm) and the *yabunikkei* or Cinnamomum japonicum (a form of wild cinnamon). Direct access to the grove is forbidden.

In Thailand, sacred groves, mostly connected to Thai folk belief, are known to have existed since medieval times. Recently, new areas are being marked off as sacred grove as an environmental movement.

Malaysia boasts to having 19 tribes; 18 tribes of Orang Asli; and one tribe called the Malays. Tree reverence is an ancient practice among the tribes and the practice is being followed also at present. The tribes' practice of tree planting around houses in such a method that the walls and wooden structures are allowed to give way to the roots of creeping plants. In recent times with increased migration towards urban centres, this wood, bamboo, woven palm leaves-houses are abandoned and allowed to return to nature. Besides, they practice of creating arches of vine and creeping flowering plants so that each time one were to enter the gates of the house, the person has to bow, giving respect upon entry to a sacred grove which were practiced by their ancestors. Such practices are even performed by people who have migrated into the cities. A garden of fruit trees surrounded by larger trees are planted around the houses to provide shade and an illusion of being at 'home' as well as to provide sustenance (in the form of fruits and seeds) to small mammals, birds and insects. Commonly, many cats are kept by the house owners to patrol the grove and guard from harmful spirits and rats which are believed to carry unclean spirits and diseases.

Tree reverence among the Tribes can be seen in the *graveyards* which are considered as holy place, on which no stone structure can be built upon. The graveyards are covered by large and tall trees, with very dense canopy that the scorching sun rays do not hit the ground and the temperatures drop making the area comfortable cool. There is also a ritual of planting small tree sapling on fresh graves by family members who will then water it and tend to it periodically. The Malays regard visiting the graves from between sunset to sunrise as it is believed that as sunrise is the beginning of day to mankind, sunset is perceived as the beginning of *day* to those who have died and dwell in the grave area. However, women and children are not allowed to visit the graveyards during night. (based on en.wikipedia.org/wiki/sacred_groves_of-India)

3. Classification of Sacred Groves

In India sacred groves are usually classified under three categories:

(i) Traditional Sacred Groves – It is the place where the village deity resides, who is represented by a symbol?

(ii) Temple Groves –a grove is created around a temple and conserved.

(iii) Cremation Ground Groves- groves around the burial or cremation grounds of villages.

4. Distribution of Sacred Groves in India

In India, sacred groves are scattered all over the country, and occur in a variety of places - from scrub forests in the Thar Desert of Rajasthan maintained by the Bishnois, deciduous forests in Easternghats, maintained by local tribes, Odias and Telugus, to rain forests in the Kerala and Karnataka of the Western Ghats. Himachal Pradesh in the North and Kerala in the South are specifically known for their large numbers of sacred groves. The Kodavas of Karnataka maintained over large number of sacred groves in Kodagu alone. In Odisha, sacred groves are scattered all over the state and many of them are protected by tribes. These groves do not enjoy protection by law, especially by the environmental laws enacted by the Parliament of the country or by any state legislation. Time has come to review this matter; Most of these sacred deities are associated with local Hindu gods. Sacred groves of Buddhist and Islamic prayer places are also known. There has been no comprehensive study on the sacred groves of the entire country.

Table 1. State-wise distribution of sacred groves in India

State	Number of Sacred Groves	Names of Important Sacred groves
Arunachal Pradesh	65	*Gumpa* Forests (Sacred Groves attached to Buddhist monasteries)
Andhra Pradesh (Simandhra & Telengana)	750	
Assam	40	Madaico, Than
Chhattisgarh	600	Sarna, Devlas, Mandar, Budhadev
Gujarat	29	
Haryana	248	
Himachal Pradesh	5,000	*Deo Bhumi*
Jharkhand	21	Sarana
Karnataka	1424	*Devara Kadu*
Kerala	2000	Kavus
Madhya Pradesh	21	*Devkot, Matikot, Devsthali, Budhadev*
Maharashtra	1600	Devrais
Manipur	365	*Gamkhap, Mauhak* (sacred bamboo reserves)
Meghalaya	83	*Law Lyngdhoh*
Odisha	322 in Similiguda, Koraput district, 81 in Sundergarh district, 20 in Sambalpur and Jharsuguda districts. Detail data from other districts not available. However the sacred grove of 'Nrusingnath' of Bolangir district, Harishankar of Bargarh district, 'Kapilas' of Denkanal district,Tara-Tarini grove of Ganjam district are well known throughout Odisha.	*Jahera, Thakuramma* Ravishankar, Nrusinghnath,Kapilas, Tara-Tarini,Andhari,Kalpeswar
Pondicherry	108	*Kovil Kadu*
Rajasthan	09	*Orans, Kenkris, Jogmaya*
Sikkim	56	*Gumpa* Forests
Tamilnadu	503	*Kovil Kadu*
Uttaranchal	18	*Deo Bhumi, Bugyal* (sacred alpine meadows)
West Bengal	670	*Garamthan, Harithan, Jahera, Sabitrithan, Santalburithan*

Table 2.Sacred groves (SG) in some districts of Odisha

Sl. No.	Name of SG	District	Approx. Area	Dominant plant species	Management pattern	Tribal group involved in the worship	Deity
1	Medha	Sundergarh	23ha	*Shorea robusta* Roth	Temple trust	Dehri, Munda	Shiva
2	Papanga	Bargarh	200ha	*Cleistanthus collinus* Roxb.	Temple trust, vana Surakhya Samiti(vss), forest	Gond, Bhil, Bhuyan	Budharaja (Shiva)
3	Andhari	Jharsuguda	1000ha	*Shorea robusta*	vss	Gond	Andhari
4	Dedungri	Sambalpur	30ha	*Pterocarpus marsupium* Roxb.	Village	Gond	Gosein baba (Shiva)
5	Gugarpat	Sambalpur	0.8ha	*Shorea robusta* Roth	Community	Gond	Gugarpaat
6	Pitabali	Bargarh	0.5ha	*Sterbulus asper* Lour.	Community	Gond	Pitabali
7	Demul (lakarma)	Sonepur	0.5ha	*Anogeissus latifolia* (DC.) Wallich *ex* Guill. & Perr.	Community	Kandha	Kandrapat
8	Mukteswar	Sambalpur	25ha	*Shorea robusta*	Temple trust, vss	Gond, Dehri	Shiva

Sl. No.	Name of SG	District	Approx. Area	Dominant plant species	Management pattern	Tribal group involved in the worship	Deity
9	Kalpeswar	Sambalpur	0.5ha	*Terminalia arjuna* Roxb.	Community	Dehri	Shiva
10	Tabdabahal	Sambalpur	1000ha	*Shorea robusta*	Village, forest	Gond	Ganpaat (Sri Ganesh)
11	Durgeikhol	Bargarh	200ha	*Cleistanthus collinus* Roxb.	Village	Gond	Durga (Goddess)
12	Padhanpat	Deogarh		*Shorea robusta*	Forest	Kandha	Padhanibudhi9 Goddess)
13	Rambhadevi	Deogarh	20ha	*Shorea robusta*	Village	Kandha	Rambha Devi (Goddess)
14	Demul (Larasara)	Sambalpur	0.2ha	*Sterbulus asper* Lour.	Community	Gond	Gram Devi (Goddess)
15	Budharaja	Sambalpur	30ha	*Cleistanthus collinus* Roxb.	Vss	Dehri	Budharaja (Shiva)
16	Bendalpat	Bargarh	30ha	*Shorea robusta*	Vss	Gond	Bendalpat

Sl. No.	Name of SG	District	Approx. Area	Dominant plant species	Management pattern	Tribal group involved in the worship	Deity
17	Demul (banpali)	Sundargarh	0.5ha	*Butea monosperma* Lam.	Community	Gond	Banpat
18	Bandurga	Sambalpur	2ha	*Shorea robusta*	Vss	Common	Ban Durga (Goddess)
19	Bindhyabasini	Bargarh	0.8ha	*Cleistanthus collinus* Roxb.	Vss	Common	Bindhabasini(Goddess)
20	Punjipathar	Bolangir	200ha	*Cleistanthus collinus* Roxb.	Vss	Gond	Dangarbudha (Shiva)

Note: The deities are lord Shiva or Goddess Durga, although names vary depending upon the locality. The tribes are Gond, Dehri, Munda, Bhil, Bhuyan and common people. The name of the SG is given usually after the deity.

Around 13, 270+ sacred groves have so far been reported from all over India, which act as reservoirs of rare flora, and rare fauna, amid rural and even urban settings. Experts believe that the total number of sacred groves in India could be as high as one lakh (100, 000).

Sacred Groves in Odisha

Table 2 gives the sacred groves of few districts of Odisha.This list is in addition to the already listed sacred groves (322) in Table 1. The table lists the name of the sacred grove, the dominant plant; the tribe/community worships the deity and the name of the deity.

Figure 2. Religious offerings in Andhari sacred grove in Jharsuguda district, Odisha.

Figure 2. Religious offerings in Andhari sacred grove in Jharsuguda district, Odisha

Ecological Significance of Sacred Groves

(i) **Biodiversity Conservation** – The sacred groves are important biodiversity Repositories of flora and fauna that have been conserved by local Communities in a sustainable manner. They are often the last refuge of Endemic species/rare species and medicinal plants in the geographical region/ landscape.

(ii) **Recharge of aquifers** – The groves is usually associated with ponds, streams or springs, which help meet the water requirements of the local communities. These aquatic ecosystems act as ground water rechargers/rain water harvesters. The vegetative cover help making the place cool because of transpiration by the plants and help in the recharging the aquifers.

(iii) **Soil conservation** - The vegetation cover of the sacred groves protect soil from erosion and brings soil stability of the area.

(iv) **Aesthetic, and other Values**- The traditional and temple sacred groves create an aesthetic atmosphere for recreation by the local community (Figure 2) and have a tourism potential. Sacred groves are also places of social gathering on special occasions, religious gatherings on special occasions (Figure) and have social integration value. In recent times, in some areas Vanamahochhav is also organized in these groves.

(vi) **Tool for Environmental Education**- The students of the area take advantage of the presence of sacred groves to learn plant taxonomy, faunal associations, spatial and trophic niche segregation of biodiversity, food chain relationships etc.

Threats to the sacred groves

Sacred groves in many parts of our country have been destroyed due to many reasons.

The common threats identified are:

(i) Pressures due to increasing livestock, fuel wood, medicinal herb collection, silting of water resources etc. **(ii)** Rapid urbanization and developmental interventions such as roads, electric lines, railways tracks, dams including commercial forestry and encroachment has led to the shrinkage of some of the largest groves in the country. **(iii)** Cultural invasion and Disappearance of the traditional belief systems, which were fundamental to the concept of sacred groves. These systems and their rituals are now considered mere superstition.

(iii) '**Sanskritisation**'- Many groves are suffering due to the transformation of the forms of nature worship into formal temple-idol worship outside the groves, known as 'Sanskritisation'. New Temple (s) and Shrines construction in the sacred groves also destroys groves.

(iv) **Invasion by exotic weeds** such as *Eupatorium odoratum, Lantana camara and Prosopis juliflora* is a serious threat to some groves.

(v) Plantation of commercial species of plants including with exotic species.

However, recently steps are being taken to catalogue the existing sacred groves, to carry out research and to conserve them.

Acknowledgement

Dr.Satyendra Prasad Mishra and Prof.Niranjan Behera of Sambalpur University provided the photographs and data on Odisha given in Table 2.. Prof.Dhruba Raj Naik showed his interest to publish this article in his edited magazine 'Andhari'. I wish to thank them profusely.

References

A guide to Japan's sacred forests at *kateigaho.com*

Asato Susumu, From Gusuku to Utaki : Okinawa's Sacred Areas from an Archeological Perspective, Board of Education of Urasoe

Boakye Amoako-Atta, Preservation of Sacred Groves in Ghana: Esukawkaw Forest Reserve and its Anweam Sacred Grove, Working Papers, South-South Co-operation Programme for Environmentally Sound Socio-Economic Development in the Humid Tropics, UNESCO

C. Dorm-Adzobu, O. Ampadu-Agyei, and P. Veit; Religious Beliefs and Environmental Protection: The Malshegu Sacred Grove in Northern Ghana; World Resources Institute and African Centre for Technology Studies, Washington DC, 1991

Down to Earth on sacred groves- A series of articles in the journal

Entry at the UNESCO website

Entry at the United Nations Division of Sustainable Development

Herodotus, v.119. Herodotus adds that "the Carians are indeed the only people we know of to conduct sacrifices to Zeus Stratios"; the connection of the presiding deity at Labraunda to Hellene Zeus is simply *interpretatio graeca*.

Horace L. Jones, (Ed.). The Geography of Strabo. Vols 1-8, containing Books 1-17. Harvard University Press and Heinemann, 1917-32

http:// en.wikipedia.org/wiki/sacred_groves_of-India).

James Frazer, The Golden Bough, Dover reprint of 1922 abridged edition, (ISBN 0-486- 42492-8)

M. D. Subash Chandran Madhav Gadgil, 2014. "Sacred Groves and Sacred Trees of Uttara Kannada"

Malhotra, K. C., Ghokhale, Y., Chatterjee, S. and Srivastava, S., Cultural and Ecological Dimensions of Sacred Groves in India, INSA, New Delhi, 2001

Malhotra, K.C., Gokhale, Y., Chatterjee, S., Srivastava, S. 2001.Cultural and Ecological Dimensions of Sacred Groves in India, Indian National Science Academy, New Delhi.

Maurice Meuleau, *Les Celtes en Europe*, Éditions Édilarge, Rennes, 2004, (ISBN 2-7028-9095-4)

Michael O'Neal Campbell, Traditional forest protection and woodlots in the coastal savannah of Ghana, Environmental Conservation (2004), 31: 225-232 Cambridge University Press

National Archeological Museum of Spoleto website entry for the exhibit of the inscribed stones

Pandey, D.N. 1998. *Ethno forestry: Local Knowledge For Sustainable Forestry And Livelihood Security.* Himanshu publications, Udaipur

Ramachandra Guha, The Unquiet Woods, University of California Press, 2000 (ISBN 978-0520222359)

Ronald Hutton, *The Druids*, p97 ISBN 978-1-85285-533-8

Ronald Hutton, *The Druids*, p97-8 ISBN 978-1-85285-533-8

Secretariat of the Convention on Biodiversity: "Protected Areas in Today's World: Their Values and Benefits for the Welfare of the Planet, " CBC technical Series No. 36.

Shimogamo-jinja: "Tadasu-no-mori (Forest of justice) "

Taylor, Isaac (1864). *Words and places: or, etymological illustrations of history, ethnology, and geography*. New York: Macmillan. OCLC 13735328.

Terry, Philip. (1914). *Terry's Japanese empire*, p. 479.

Tshan, Francis J. Adam of Bremen

Venceslas Kruta, *Les Celtes, Histoire et dictionnaire*, Robert Laffont, coll. Bouquins, Paris, 2000, (ISBN 2-7028-6261-6)

Wikimedia Commons has media related to ***sacred groves***

Wonder Okinawa: World Heritage site inclusion

5

Soil Biodiversity and Earthworms

'It may be doubted whether there are any other animals which have played so important a part in the history of the world, as have these lowly creatures' Charles Darwin, 1881.

Introduction

Tropical soil biology fertility institute in Nairobi, an international centre for tropical agriculture has united 300 scientists and is running research projects in seven countries including India to survey soil organisms since 2002 and aim to identify living indicators for soil fertility, of poor soils, and has already identified some novel organisms, that are useful. Soil stockpiles about 1500 Gigatonnes of organic carbon, more than earth's atmosphere and all the plants (FAO).

If the soil remain degraded and many soil biodiversity disappear, the planet might lose access to beneficial organisms that enhance crop yield, detoxify toxins and make useful by-products such as antibiotics, used in pharmaceutical industries before they are even discovered.

Soil harbours bacteria, symbiotic and non-symbiotic nitrogen fixing bacteria, actinomycetes (source of antibiotics), viruses, fungus including mycorrhiza (fungus roots), algae, protozoa, nematodes, enchytraeid and earthworms, arthropods, mollusks and many others. Since Charles Darwin and his contemporaries' time, it is known that it is the role of these organisms as biological ploughs and suppliers of key nutrients for plant growth. Actinomycetes are numerous and many of them are antibiotic producers. Fungi occur mainly in surface soils of wide range of habitats under many land uses and vegetation. Many species develop a symbiotic relationship called mycorrhiza or fungus roots which provide water and nutrients to the plants and are indicators of fertile soils. The soil biota and their influences in soil are listed in Figure 1 and 2 and Table 1.Fungal strands of *glomus, acaulospora, gigaspora*, and *sclerocystis (Figure 3, 4)* are common associations with crop roots. Ectomycorrhiza are also common association with angiosperm and gymnosperm tree roots.

The soil fauna contributes maximum biomass in tropical soils. Earthworms and termites play major role in influencing decomposition, nutrient release in soil. Earthworms are used as active agents in Vermitechnology for waste land reclamation, plant growth and composting. The wealth of new species in soil waiting discovery, especially in tropics. They represent huge new genetic sources which will yield an array of antibiotics, drugs.

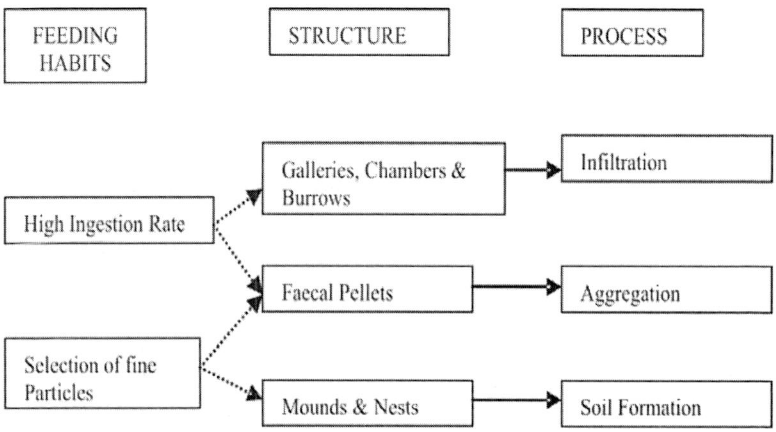

Figure 1. Relationships Between Soil Invertebrate Behavior And Soil Physical Process

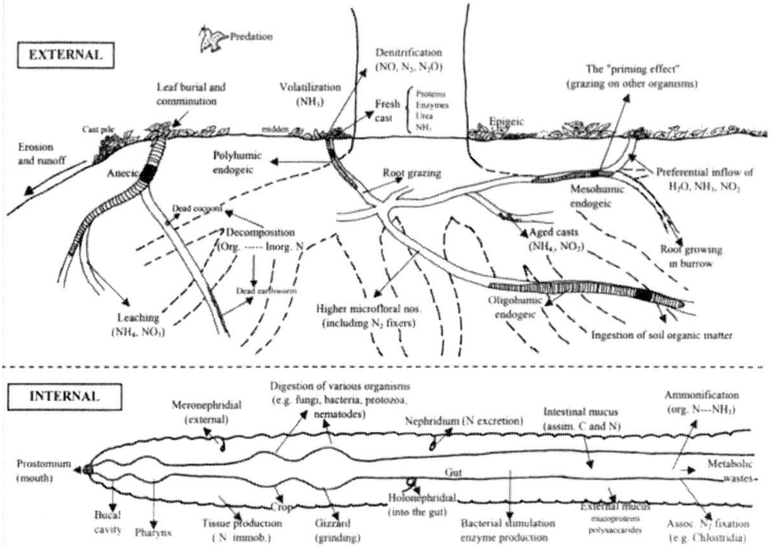

Figure 2. Schematic representation of the functional relationships between earthworms and their internal and external environment (Doube and Brown, 1988)

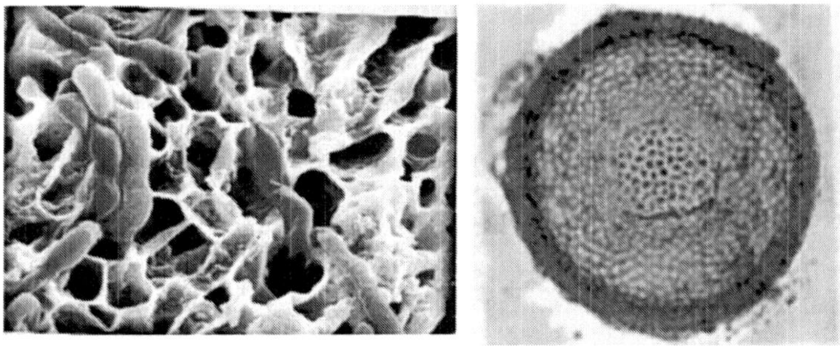

Figure 3. Mycorrizha Fungal Spores

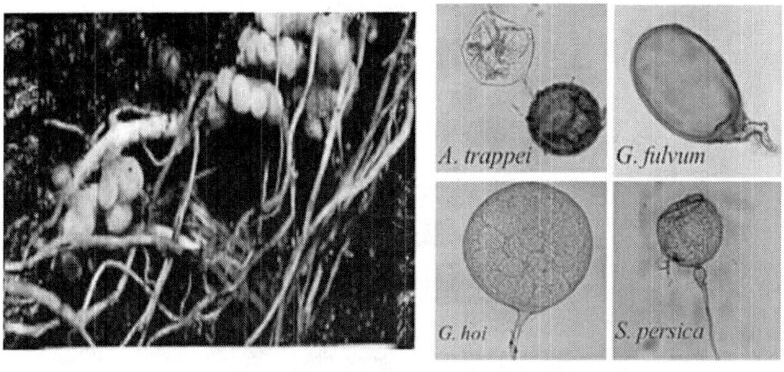

Nitrogen fixing bacteria (nodules)

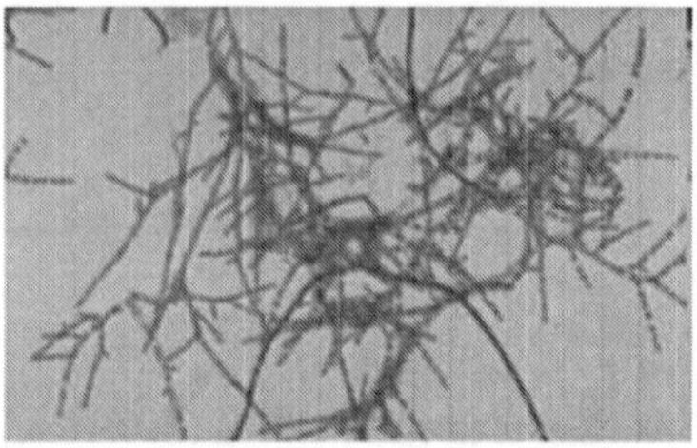

Bacterial spores Actinomycetes

Figure 4

Fungi occur mainly in surface soils of wide range of habitats under many land uses and vegetation. Many species develop a symbiotic relationship called mycorrhiza or fungus roots which provide water and nutrients to the plants and are indicators of fertile soils. Fungal strands of glomus, acaulospora, gigaspora, and sclerocystis are common associations with crop roots. Ectomycorrhiza are common association with angiosperm and gymnosperm tree roots.

Table 1. Influences of Soil biota

Micro flora	Catabolism organic matter, mineralize and immobilize nutrient in aggregates	Produce organic compounds that binds agreements, hyphae entangle particles unto aggregates
Micro fauna	Regulate bacterial and fungal populations after nutrient turnover	May affect aggregates structure though interactions with micro flora
Meso fauna	Regulate fungal and micro faunal populations, After nutrient turnover, fragment plant residues	Produce fecal pellets, create biopores, promote humificationcal pellets
Macro fauna	Fragment plant residues, Stimulate microbial activity, redistribute organic matter and microorganisms	Mix organic and mineral particles, create biopores, promote humification, produce fecal pellets

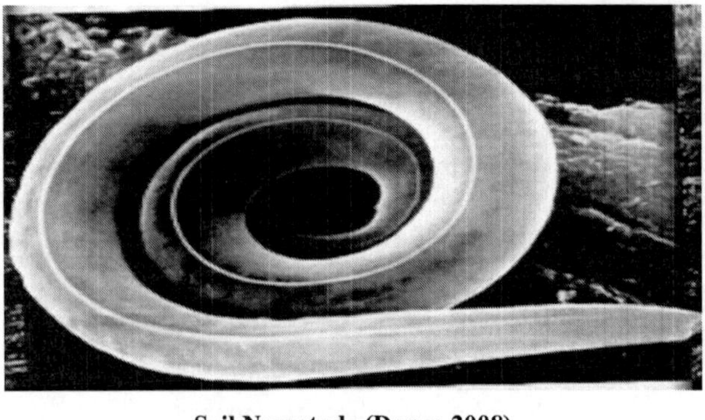

Soil Nematode (Dance 2008)

Table 2. Soil Faunal Biodiversity

Taxa	Total No. of Species		Endemic Species	Geographical distribution
	(World)	(India)		
Echiura	127	43	12	Coastal areas
Chilopoda	3000	156	NA	South Indian, N.E India,
Diplopoda	7500	162	NA	Mumbai, Gujarat, Bihar, UP, HP, & Orissa
Orthopera	17250	1750	200	Almost entire India
Dermoptera	2000	320	-	Almost all over India
Embioptera	200	33	14	Discontinuous distribution, Himalayas, Gangetic plains, Barkuda island, eastern ghats, Western Ghats
Isopteran	2000	253	170	Almost entire India
Hemiptera	NA	6500	1335	Restricted distribution, Eastern Himalayas, Western Himalayas, Peninsular India
Neuroptera	5000	335	262	Himalaya, Desert, peninsular India, Insular ecosystem
Coleoptera	800000	15500	3113	Wide spread distribution
Mecoptera	350	15	-	Eastern India, UP, Maharashtra
Diptera	100000	6093	2132	Widespread
Trichoptera	7000	812	649	Himalayan & Peninsular India
Annelida	12000	1000	-	Wide spread

Source: Dr.J.R.B.Alfred-Zoological survey of India, Kolkata

Soil organisms have key roles in modern agriculture

Soil scientists are now convinced that soil biodiversity, their operational mechanism and interaction among themselves & with soil is the key to restoring the fertility of degraded lands and to increase crop yields in tropics without use of heavy dose of fertilizers and pesticides.

Soil biologists are usually of the opinion that there is no environment on the planet that is biologically more diverse per unit area than soil. Recent studies on DNA sequencing of soil biota also indicate such a situation (Dance, 2008,). Looking at example set by the soil biology researchers of Sambalpur University in India, (since 1973-), a team of experts from French Institute De Recherché Pour le Development and the School of life sciences, Sambalpur University, Odisha, India, worked since 1981 with the Indian Company Parry Agro, and made trials at several 100 years old tea plantations in the state of

Tamil Nadu, where tea yields have stalled despite heavy use of fertilizer and spraying of plant growth hormones and pesticides, indicates that asserting the role of below ground biodiversity can have huge agricultural benefits.

Soil harbours symbiotic and non-symbiotic nitrogen fixing bacteria, other bacteria, actinomycetes (source of antibiotics), and viruses, fungi including fungal roots (mycorrhiza), protozoa, and algae. The wealth of new species in soil waiting discovery, especially in tropics. They represent huge new genetic sources which will yield an array of antibiotics, drugs. In July 2008, United States Research Council listed soil quality as the biggest barrier to higher crop yields for farmers in South Asia and Sub-Saharan Africa. They are of the opinion that understanding the interaction of soil organisms is crucial to creating healthy productive ecosystems. *Recent trends are that since soil biodiversity is diverse and rich; DNA sequencing in any soil will help to understand their diversity and functioning. It will help soil management for agriculture and forest & horticulture management.*

Soil Fauna: The Earthworms

Aristotle (330 BC) called earthworms as the **'intestines of the earth'** indicating their importance in thc progress of human civilization. He called them intestines of the earth as he observed that they ingested huge amount of soil.

Darwin's Plough: Charles Darwin (1809-1882) was born on 12 February 1809. Scientists all over the world celebrated the bicentenary birth year in 2009 of this great scientist whose discovery formed the foundation of modern evolutionary theory and also of soil biology. In 1836, Darwin went in his voyage on the Beagle. In 1837, one year after his voyage on the Beagle, he paid a visit to his maternal uncle Josiah Wedgwood, (later he became his father-in-law), who took him to many fields, where 15 years before, the soil surface had been covered with lime and partly burnt wood/coal. These materials were now buried at deeper layers of soil and Josiah Wedgwood thought that earthworms did it (Desmond and Moore, 1992).This observation and comment by his uncle triggered Darwin's scientific curiosity to study them and to observe their activity.

Charles Darwin in 1881, six months before his death wrote a book entitled 'The formation of Vegetable Mould through the Action of

Worms with observations on their Habits'. The book was a "best seller" as 3500 books were sold in the first month of its publication. The book showed the importance of earthworms in burial of surface lying organic matter and opened a new line of research in Soil Biology. Darwin was convinced by his study and observations that the subterranean biota, especially earthworms played crucial role in determining ecosystem functions. He later commented that earthworms played a beneficial role in soil formation and aided soil fertility and stated "worms have played a more important role in the history of the world than most persons would at first suppose". Earthworms are fondly called Darwin's Plough. Later many scientists in Europe also observed the importance of earthworms in soil in the formation of worm-casts and humus. Earthworms are helpful agents in the advancement of human civilization by improving quality of earth's soil and linked to plant productivity. Charles Darwin wrote in his book that the plough is one of the most ancient and most valuable inventions; but long before man existed, the land was in fact regularly ploughed, and still continued to be thus ploughed, by earthworms. It may be doubted whether there are any other animals which have played so important a part in the history of the world, as have these lowly organized creatures'.

Earthworms contribute > 80% of soil living organism's biomass in many ecosystems. (Dash, 1978, Dash, 1990). Earthworms and termites play major role in influencing decomposition, nutrient release in soil. Earthworms are used as active agents in vermitechnology for waste land reclamation, land use management, plant growth and composting. J.M. Julka is the leading earthworm taxonomist of India. Nine families, 69 genera and 418 species including 44 exotic species have been reported till 2012 from India.

Successful Vermitechnology trials were made at Hindalco Industries, Hirakud, Odisha in 1980s and many other places, to generate value added products, like organic manures, animal feed, materials using municipal, canteen etc., and organic wastes. The studies indicated huge organic pollution abatement potential by the earthworms. However their functioning is intriguing the scientists. This has triggered the new UNEP / TSBF project in many parts of the world including India.

Earthworm-Microorganism Interaction

M.C.Dash and his associates pioneered Oligochaeta-microorganism interaction studies in India. Many studies (Dash and Cragg, 1972, Dash et al, 1979, 1980) show that earthworm and nematode, earthworm and protozoa, earthworm-soil micro fungi interactions in soil play important role in soil activities and help controlling nematode & protozoa populations. Dash et al (1980) in a pioneering work designed experiments to study feeding of nematodes by *Lampito mauritii* earthworms and found that the nematode populations due to earthworm predation decreased about 21 to about 54 % of their population. Gut content analysis showed that the earthworms fed mainly on non-parasitic nematodes, as their population decreased by maximum of 90% compared to maximum of 47% decrease of parasitic nematodes. Earthworms graze over and digest microorganisms, especially soil fungi and their castings act as foci for dissemination of some species of soil fungi. In recent studies, Behera, Mishra and others have done earthworm-micro flora interaction studies in fly ash mund reclamation sites.

Agro-pesticides and Earthworms

M.C.Dash and his associates also pioneered studies in India on the effect of agro-pesticides on earthworms. In modern intensive agricultural systems, organochlorine, organophosphates, carbamate and metal containing chemicals are used to control pests. Although earthworms are non-targeted animals, they are affected by these chemicals. Laboratory studies show that LC 50 values based on log dose vs probit mortality follow the order of *Lampito mauritii* > *Drawida willsi* >*Drawida calebi* >*Octochaetona surensis* (all epigeic or anecic species) for tolerance to monocrotophos; *Lampito mauritii* > *Drawida calebi* >*Octochaetona surensis* > *Drawida willsi* for tolerance to fenitrothion (Patnaik and Dash 1990). Worms become ureotelic and ammonotelic due to starvation with increasing trend of excretion when exposed to monocrotophos (Patnaik and Dash 1991). Cholinesterase activity of nerve and muscle tissues of *Lampito mauritii* and *Drawida calebi* were inhibited in both short term and continuous exposure to monocrotophos and fenitrothion. The recovery potential varied by species (Patnaik and Dash 1992 and 1993). Panda and Sahu (1999) studied effects of Malathion on the growth and reproduction of *Drawida willsi* under laboratory conditions. The growth was inhibited in the first two weeks of application of the pesticide but worms recovered and normal growth and reproduction occurred after 105 days of application of Malathion. Comparative toxicity data based on the effect of agrochemicals on Indian earthworms are not available from different areas enabling identification of suitable test species and confirmation of conclusions based on laboratory studies. *Lampito mauritii*, because of its common occurrence, will be suitable as the test species of choice for toxicity studies.

Earthworms immobilize and also excrete cadmium, cobalt, mercury, zinc, copper and lead in their tissues (Beyer and Cromartle1978, Beyer et al.1987, Marinussen et al.1997, Panda et al. 1999, Nahmani et al. 2003). Panda et al. (1999) worked on the accumulation and effect of zinc on the growth, reproduction and life cycle of *Drawida willsi* and found that earthworms were able to regulate their body content of zinc within a range of 116-125 mg kg^{-1} (dry weight). Reproduction and completion of life cycle was significantly affected negatively when Zn concentration in soil exceeded 200 mg kg^{-1}. Nahmani et al. (2003) worked in 11 sites including grasslands, cultivated soils and poplar plantations varying in the level of Zn, Cd, and Pb contamination and found that soil pollution reduced density as well as species richness of earthworms. *Aporrectodea caliginous* and *Allolobophora chlorotica* appeared to be most sensitive to heavy metals. Although comprehensive data base on the effect of pesticides, and heavy metals on earthworms are yet to develop, there are evidences to show that earthworms tend to overcome the stress by increasing mucous secretion, reducing borrowing activity and increasing reproduction.

Earthworms, Energy Flow and Carbon Sequestration

M.C. Dash pioneered energy flow studies in earthworm populations. In many ecosystems earthworms play most important role in the energy flow in the system (Dash and Patra, 1977, Dash et al, 1979, Senapati and Dash, 1981, Dash and Dash, 2008, Dash et al, 2009, Dash, 2012).

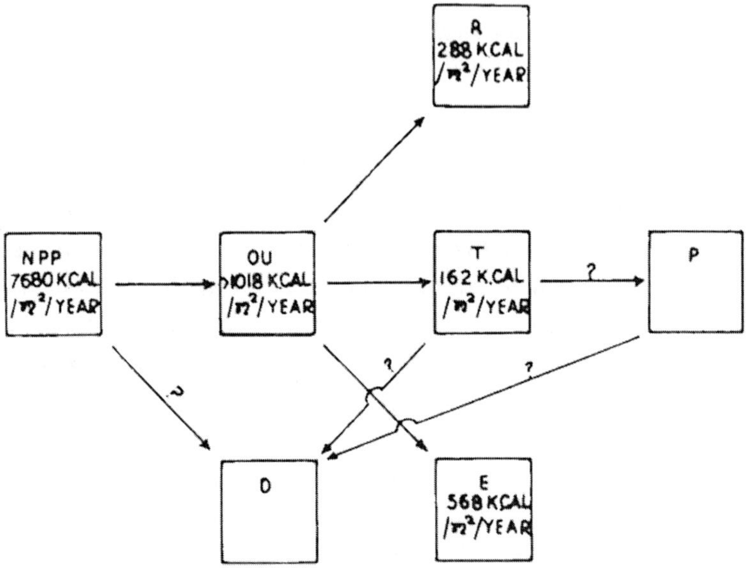

Energy flow studies in Pastures in Orissa (Dash and Patra, 1977)

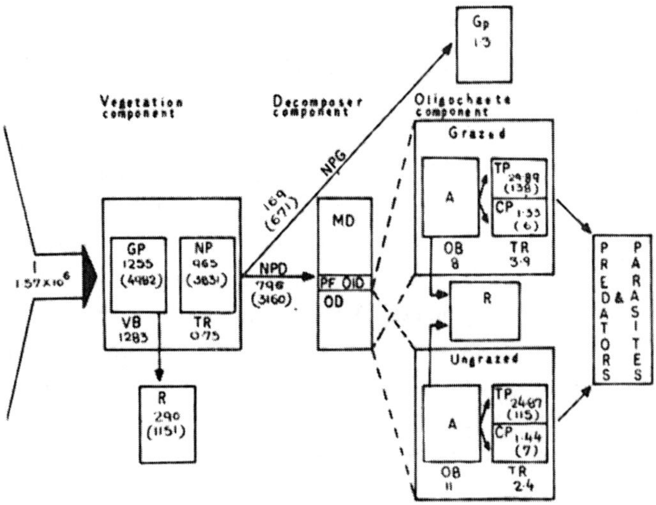

Energy flow studies (Senapati and Dash, 1981)

Energy flow studies indicate that earthworms play key functional role in soil system, soil fertility and in carbon sequestration.

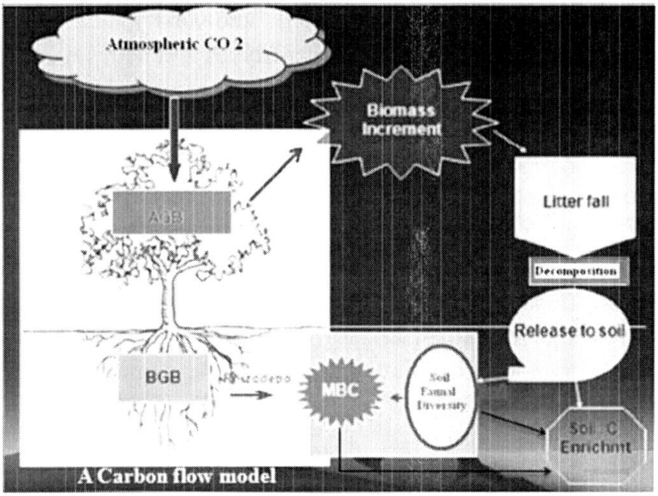

Carbon sequestration process involving soil biodiversity (Dash, 2013, Dash, Behera, 2013)

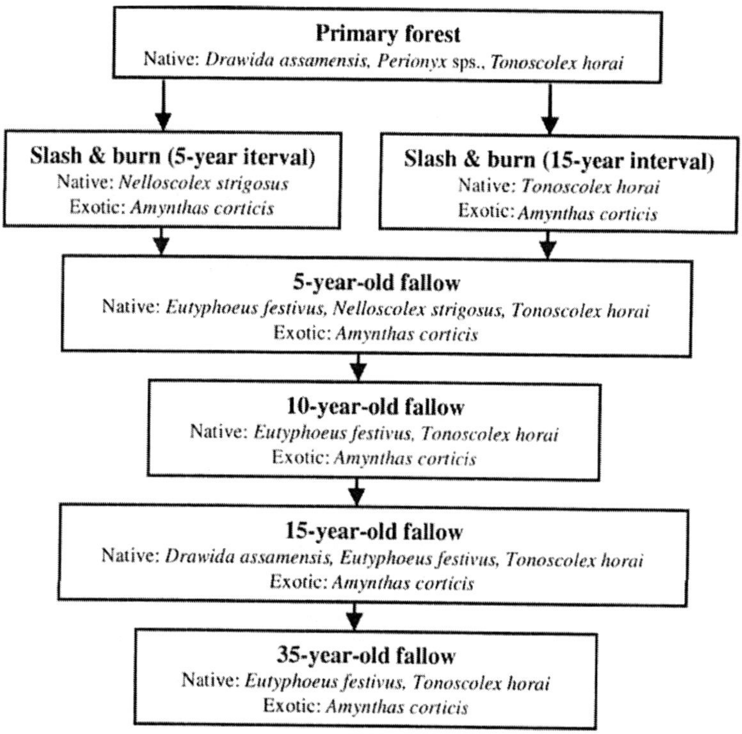

Earthworms and Land Use Management

Soil scientists are convinced that soil biodiversity, their operational mechanism and interaction among themselves and with soil is the key to restoring the fertility of degraded lands and to increase crop yields in tropics without use of heavy dose of fertilizers and pesticides. Studying earthworm population, biomass and turn over in different land uses provide insight to manage degraded lands and productive lands etc. Earlier studies carried out at Sambalpur and more recent and elaborate studies carried out by Ramakrishnan, Tunira Bhadauria and others in north-eastern region, in Himalayan ecosystems by Maikhuri, Saxena, Rao, Bhadauria and others, in rubber plantations by P.S.Choudhury and others have opened up a new line of soil biology research in India.

Earthworms for Vermitechnology

On the basis of the pioneering field and laboratory studies in India, Dash and Patra (1977 and 1979), Dash (1978), Dash and Senapati (1982), Dash et al (1985, 1986) suggested to utilize the native species of earthworms like *Lampito mauritii*, *Drawida willsi*, *Drawida calebi and Octochaetona surensis, Perionyx excavatus* and the exotic species *Drawida bolaui*, which has been successfully naturalized in India to be used in vermicomposting and for their mass culturing. In two recent reviews on the role of earthworms in tropical ecosystems, especially in different land uses in India, Dash and Dash (2008,) and Dash et al (2009) have suggested using them for Vermitechnology in wasteland reclamation, crop productivity and compost-manure (vermicomposting) preparation on a large scale.

Large scale successful field applications of vermicomposting have been made by Radha D.Kale in Karnataka. Bikram Senapati, Patrick Lavelle and others have patented a technology of using earthworms and soil biota in tea gardens of south India to increase tea production. Vikram Reddy and his group have used earthworms for ameliorating activity in soils contaminated with heavy metals.

Questions raised on earthworms

Despite these pioneering efforts and advances in earthworm research made in other centres of India, (Radha D.Kale, M.V. Reddy, P.S.Choudhury, Janardan Singh, and others), many scientists and people are not aware what an earthworm does for their soil and what are the biological and ecological requirements of these largely helpful worms ?. How many species of earthworms are found in India, or in different provinces of India and in the World? In many awareness, training programmes, and seminars, I participated in different parts of India relating to their biology, ecology, Vermiculture (earthworm culture), and Vermitechnology and observed that although farmers are aware of their beneficial role in agriculture but they do not know what the worms do precisely.

Our school and college students often ask questions on them: How many types of earthworms are there? How big earthworms can grow? How long they live? What is the food of earthworms? How to identify them? How earthworms move in soil? How do they breathe? How do they reproduce? How to collect and study them? How to preserve them? Which are the species good for vermicomposting? How to design research work on them? And many such questions arise in their mind.

Soil is the abode of many species of microorganisms and soil fauna. Among the soil fauna, earthworms form a major component and dominate the soil fauna in many land uses.

Vermicomposting

In recent years earthworms have gained much importance because of production of vermicomposting (a rich manure) through them and also their utilization as a source of animal protein for poultry, and fish, and utilizing them in wasteland reclamation, crop productivity and utilization of their castings as water retaining material in the preparation of bed for growth of edible mushroom and for other uses. It has been estimated that few lakh farmers in India practice vermicomposting. India having 418 species (4000 in the world) till date is considered rich in earthworm fauna..

Perionyx excavatus

(Clitellum close to head; colour uniform red or bluish, male pores seg 18)

Indian vermicomposting worm

Earthworms in Medicine and food

Chinese have used earthworms as medicine to cure human diseases like piles, sore throat, chronic cough and bronchitis, jaundice, impotency and some other diseases. There is report that folk medicines in Burma and China use earthworm extracts to treat pyorrhea, antipyretic activity, rheumatism etc. Chinese and scientists in Far East countries have extracted peptides and some lipids and archiclonic acid from their tissues and have utilized them in the preparation of medicines. It has also been reported that Chinese and Japanese scientists have isolated Lumbrokinase enzyme from earthworm body and prepared capsules for treatment and prevention of heart diseases.

In Burma, the use of earthworms in the treatment of a disease called ye se kun byo, the symptoms of which suggest pyorrhea is practiced since 1, 340 A.D. (Stevenson, 1930, Reynolds and Reynolds, 1972).The worms are heated in a closed pot until reduced to ashes, which are used as a tooth powder or better palatability, are combined with roasted tamarind seeds and betel nuts.In another disease, mainma meephwanoyeekhun thwaykhan, a woman feels postpartal weakness and finds difficult to nurse her infant. Worms are boiled in water with salt and onion and the clear fluid is decanted and mixed with patient's food (Gates, 1926).It has been reported (Bristowf (, 1932) that earthworms were used to treat smallpox in Laos, and Burma. Earthworms are soaked in water the patient is bathed in the liquid. Patient is given to drink powder prepared from roasted worms mixed with coconut water. The treatment hastens the severity of the disease and reduced mortality from 100 percent to 25 percent (Reynolds and Reynolds, 1972).In Iran, worms are baked and eaten with bread to reduce the size of bladder stone and dried worms are eaten to cure the yellowness of a jaundiced patient. Earthworm ashes are applied to the head with rose oil to make hair grow. The Cherokee Indians of the Great Smoky Mountains of North America used worm poultices to draw out thorns. (Carr, 1951).The Nanticoke Indians of Delaware used worm-pastes to reduce rheumatic pain (Carr, 1951, Price, 1901).These are the records of folk medicines. We do not know whether the remedies cited above actually worked. This needs rigorous scientific investigations for validation.

In south East Asia, earthworm soup is considered as a delicacy and consumed because of its medicinal property. Earthworms are used as fish, poultry and animal feed and their castings are utilized in the preparation of bed to grow edible mushrooms.

The purpose of this article is to bring together the knowledge on earthworms for our school, college students and for the general educated public and for research students. This book will help biologists, ecologists, students at all levels and Non Government Organizations to get information on earthworms and generate interest in them to know more and to utilize them for mankind for organic waste management and pollution abatement, reclamation of waste land, increasing crop productivity, etc.

(Source: Charles Darwin's Plough: Tool for Vermitechnology by M.C.Dash, I.K.International Publishers, New Delhi (2012).

6

Earthworms in the Himalaya and Western Ghats Region of India: A Review

Introduction

Mountain regions of the world are fascinating as they cover a wide range of ecological diversity over smaller areas because of elevation effect and settlement of a several local communities maintaining natural ecosystems together with the managed ones, with landscape management practices varying among local communities isolated by terrain and linguistic barriers. In recent times ecotourism to mountainous regions is not only an economic activity but demands conservation of traditional as well as new ecosystems that attract tourists. Sustainable land use and resource management are key requirements for an area to get recognition as a preferred spot of ecotourism. Increasing demand for organic food and persistence of traditional organic agriculture as patches in the matrix of natural ecosystems in marginal mountain areas make them a prospective area of ecotourism. Soil biodiversity is key to sustainable organic farming (Ramkarishnan *et al.,* 2005) and earthworms are the most dominant component of soil biota in terms of biomass and crucial for maintaining soil fertility (Dash, 1978, Senapati and Dash, 1981, Julka and Paliwal, 2005; Dash *et al.,* 2009; Bhadauria *et al.,* 2012; Dash, 2012). This paper is an attempt to review the information available on diversity and functions of earthworms in the Himalaya and Western Ghats region of India, the areas distinguished globally for their highly valuable biodiversity and ecosystem services, and to identify knowledge gaps to be addressed in future research.

Distribution and diversity of earthworms: National Survey

INVENTORY India is spread over an area of 3, 287, 797 km^2 (2.4% area of the World), covering a wide range of physiographic, climatic and land use/cover types. Among the 15 agro-climatic regions identified in the country, (Fig. 1) four regions viz., Eastern Himalaya, North-east Ranges, Western Ghats and Western Himalayas figure among the global biodiversity hotspots. However, these hotspots identified based on extremely high endemic vascular plant species richness (presence of at least 0.5% or 1500 endemic vascular plant species) and vegetation degradation (loss of at least 70% primary vegetation) may not necessarily have high diversity of other plant and animal taxa (Myers, 2000; Kareiva and Marvier, 2003).

Geological histories, socio-cultural and economic conditions differ within as well as between hotspots. Thus, settled agriculture on terraced slopes constitutes the predominant agricultural land use in the western and central Himalaya and shifting/slash-burn agriculture on natural slopes in the eastern Himalaya and its extension ranges. The Himalayas is a creation of modern plate tectonic forces but not the Western Ghats, an area that has never been submerged under sea. In contrast to high-input commercial agriculture that dominates in the Indo-Gangetic plains, subsistence low-input traditional farming is widespread in the mountain regions (Bhadauria and Ramakrishnan, 2005; Senapati *et al.,* 2005; Chaudhuri *et al.,* 2008).

Zoological Survey of India has been involved with inventorying of soil fauna diversity for a long period of time, with survey efforts focused largely on presence/absence of different taxa in different environments. Nevertheless, the efforts of this organization devoted exclusively for survey and inventorying of the faunal wealth of the country are augmented by many researchers in other research and development institutions. Nine families of earthworms with 69 genera and 418 species have been reported from India. On the basis of available data, the Western Ghats and West Coastal Plains would stand out as the regions with the highest level of earthworm species richness followed by Eastern Himalayan Region, Southern Plateau, Western Himalayan Region, Eastern Coastal Plains and Eastern Ghats, Gangetic Plains, Gujarat plains, Islands, Western dry Regions and transgangetc regions (Julka and Paliwal, 2005, Dash and Dash, 2008). The Western Ghats region is home to 53% species known from India compared to 26% and 12% in the case of the Eastern Himalaya and the Western Himalaya, respectively. Drawida (38 species) is the most species rich genus followed by *Megascolex* (30 species) in the Western Ghats, *Perionyx* (33 species) followed by *Drawida* (14 species) in the Eastern Himalaya, and *Perionyx* and *Amynthas* (4 species in each Genus) in the Western Himalaya. The Western Ghats harbour 193 native species compared to 85 in the Eastern Himalaya and 22 in the Western Himalaya, though the three regions do not differ much in terms of number exotic species (25-26). Among the native peregrine species associated with agroecosystems, *Octochaetona palniensis* is confined to the Western Ghats and *Lennogaster pusillus* and *Eutyphoeus* spp. to the Western Himalaya. Among the exotics associated with agroecosystems, *Drawida japonica* was able to establish only in the Western Himalaya, while *Dichogaster affinis* and *Pontoscolex corethrurus* could establish in Eastern Himalaya and WesternGhats region but not in the Western Himalaya (Table 1).

Figure Missing in Your Article

Figure 1: Agroclimatic regions of India (from Julka and Paliwal, 2005)

Table 1: Native peregrine and exotic earthworm species associated with agroecosystems in the Himalaya and the Western Ghats region of India (based on Julka and Paliwal, 2005 and personal communication with Julka; A, absent; P, present)

Native peregrine species	Western Himalaya	Eastern Himalaya	Western Ghats
Lampito mauritii	A	P	P
Perionyx excavatus	P	P	P
Perionyx sansibaricus	P	A	P
Octochaetona beatrix	P	A	P
Octochaetona surensis			
Octochaetona palniensis	A	A	P
Lennogaster pusillus	P	A	A
Ramiella bishambari	P	A	P
Eytyphoeus incommodus	P	A	A
Eutyphoeus michaelseni	P	A	A
Eytophoeus waltoni	P	A	A
Drawida willsi	A	A	A
Drawida calebi	A	A	A
Drawida nepalensis	P	P	A
Thatonia gracilis	P	A	A
Exotic species			
Dichogaster affinis	A	P	P
Dichogaster bolaui	P	P	P
Amynthas alexandri	P	P	P
Amynthas corticis	P	P	P
Amynthas morrisi	P	P	P
Metaphire houlleti	P	P	P
Metaphire posthuma	P	P	P
Polypheiritima eleongata	A	P	P
Drawida japonica	P	A	A
Pontoscolex corethrurus	A	P	P
Ocnerodrilus occidentalis	P	A	P
Allobophora parva	P	P	P
Aporrectodea cal. trapezoiedes	P	P	P
Aporrectodearosea rosearosea	P	P	P
Eisenia fetida	P	P	P

Native peregrine species	Western Himalaya	Eastern Himalaya	Western Ghats
Octolasion tyrtaeum	P	P	P
Endemic genus/species			
Curgiona	A	A	P
Kotegeharia	A	A	P
Mallehulla	A	A	P
Priodochaeta	A	A	P
Karmiella	A	A	P
Troyia	A	A	P
Comarodrilus	A	A	P
Chaetocotoides	A	A	P
Parryodrilus	A	A	P
Dashiella	A	A	P
Moniligaster	A	A	P
Celeriella	A	A	P
Lampito	A	A	P
Travoscolides	A	A	P
Wahoscolex	A	A	P
Tonoscolex	A	P	A
Kanchuria	A	P	A
Perionyx (4 species)	P	A	A
Eutyphoeus (2 species)	P	A	A
Plutellus (1 species)	P	A	A

Earthworm communities and populations

Studies in village landscapes in central/western Himalaya (Bhadauria *et al.,* 2000, Sinha *et al.,* 2003) showed occurrence of three species in pasture soils compared to eight species in other land-uses, with Exotic *Amynthas corticis* being the most common species. Biological invasion was observed in both early and late successional stages.

Bhadauria and Ramakrishnan (2005) found three native species, viz., *Tonoscolex horai*, *Drawida assamensis* and *Perionyx* sp. in the primary forest in north-east India. Conversion of primary forest for slashes and burn agriculture resulted in the loss of two native species and colonization of the disturbed area by another native species viz., Nelloscolex *strigosus* and exotic *Amynthas corticis*. Although richness of native species increased and the native species were able to coexist with the exotic species during secondary succession after abandonment of cultivation, complete restoration of native earthworm species assemblage did not occur. A similar trend was observed in central/ western Himalayan region, though there were differences in species found in the western and eastern Himalaya (Bhadauria *et al.,* 2012).

Chaudhuri *et al.* (2008, 2009) found that conversion of primary forests to rubber plantations in Tripura led to dominance of the exotic-endogeic *Pontoscolex corethrurus* as well as a change in the functional status of this species, from endogeic in the primary forests to endo-anecic and endo-epigeic in plantations. In general the change in land use pattern or loss of primary forest followed establishment of some exotic species and also replacement of some natives by other native species.

Epigeic and anecic species, such as *Dichogaster bolaui*, *Drawida willsi*, *Perionyx excavatus*, *Perionyx sansibaricus*, *Ramiella* sp. and *Lampito mauritii* are widely distributed and valued for their use in vermitechnology in Western and Eastern Ghats (Dash and Senapati, 1985; Dash and Dash, 2008; Dash *et al.,* 2009). *Haplochetalla* spp is widespread in laterite and red soils of the Western Ghats.

Studies in different land-uses in the state of Karnataka in south India showed species composition of earthworm communities in natural forests closer to that in plantations (coffee, Acacia and cardamom plantations) and paddy fields but radically different from that in grasslands. *Pontoscolex corethrurus*, an endogeic exotic species, was found in all land-uses (Kale *et al.,* 2008, Chandrashekara *et al.,* 2008). Studies carried out in Eastern Ghat region also showed variation in species composition between land uses but a species like *Lampito mauritii* persisted in all land uses (Dash and Patra, 1977; Senapati and Dash, 1981; Mishra and Dash, 1984; Senapati *et al.,* 2005).

Population and biomass

Data available on earthworm population diversity, density, biomass in different mountainous regions of India are summarized in Table 2. Earthworm density and biomass are influenced by a whole range of abiotic and biotic factors. A high level of environmental heterogeneity and variation in land use/management practices in mountains may result in huge variation in biodiversity within an Agroclimatic region.

Metaphire anomala, *Metaphire houlleti*, *Ocnerodrilus occidentalis*, *Dendrodrilus rubidus* and *Aporrectordea calliginosa* occurred in the Nanda Devi Biosphere Reserve, a relatively cool area with lower degree of anthropogenic pressures due to legal protection but not in Hariyali landscape, a comparatively warmer area faced to more intense human disturbances in the absence of any legal protection. *Allobophora parva*, *Eutyophoeus pharapingianus*, *Octochaetona beatrix* and *Perionyx* spp. occurred in the latter area but not in the former area (Sinha et al., 2003; Maikhuri *et al.,* 2008). In areas with high intensity of agricultural land use, earthworm community comprises *Amynthas diffrigens*, *Amynthas alexandri* and *Eisenia fetida* (Kaushal and Bisht, 1994; Kaushal *et al.,* 1995), species rarely found in traditional landscapes with less intensively land uses. The latter do differ in terms of structure and composition of earthworm communities. *Bimastos parvus* and *Octolasion tyrtaeum* were sampled from a village landscape in Almora district (Bhadauria and Ramakrishnan, 2000) but not from a similar landscape in Chamoli district (Sinha *et al.,* 2003). In shifting agricultural landscapes in the Eastern Himalaya, *Drawida assamensis* and *Nelloscolex strigosus* occurred across an elevation gradient while *Megascolides astrophytes* was confined to lower elevations and *Amynthas diffringens* and *Tonoscolex horaii* to higher elevations (Mishra and Ramakrishnan, 1988; Bhadauria and Ramakrishnan, 1989). Total number of species in cultural landscapes did not vary much (6-8) but certain ecosystem types/patches may be quite poor (e.g., wet paddy fields, Cedrus forests subject to intensive disturbances) or quite rich (e.g., home gardens, rainfed agriculture) in terms of number of species (Maikhuri *et al.,* 2008; Bhadauria *et al.,* 2012).

Earthworms could be a sensitive indicator of environmental quality, e.g., presence of Ramiellona *wilsoni* is indicative of pristine tropical montane forests in Mexico (Negrete-Yankelevich *et al.,* 2007) but necessarily not in all situations. Earthworm population was not significantly influenced by intensity of management in grasslands in Ireland (Curry *et al.,* 2008) and tropical rain forest ecozone in the Western Ghats of India (Rossi and Blanchart, 2005). Density and biomass are likely to be more sensitive to environmental changes and land management practices than species richness and diversity. In the Nanda Devi Biosphere Reserve, earthworm abundance declined with decline in temperature (*i.e.,* increase in elevation) and was sensitive

Table 2: Density and biomass of earthworms (m^{-2}) in different areas in the Indian mountainous region (blank fields indicate non-availability of data)

Land use/ ecosystems	Diversity	Density (number of individuals m^{-2})	Average/ range of biomass (wet weight, g)	Trends	References
Pine woodland, Shilling, Meghalaya		9-52	4.4-9.4		Reddy and Alfred, (1978)
Mixed woodland, Solan, Himachal Pradesh		30-118			Julka and Mukherjee (1984)
Humid Tropical deciduous forest, Andhra Pradesh (Eastern Ghats)	Five Species: *Amynthas alexandri, A. diffringens, Metaphire posthuma, M. houlleti, Dichogaster* sp.	315 (28-281 in different micro sites within forest ecosystem)		Two peaks, one during and the other at the end of the rainy season	Reddy (1987)
Shifting agriculture at Nangpoh in North-Eastern India	Three species: *Megascolides antrophytes, Drawida assamensis, Nelloscolex strigosus*	68		Peak density immediately after rainy season	Mishra and Ramakrishnan (1988)
Shifting agriculture at Shillong in north-eastern India	Five Species: *Amynthas diffringens, Drawida assamensis, Eutyphoeus festivus, Nelloscolex strigosus, Tonoscolex horaii*	4-47 in cropping phase, 50 in fallow phase	-	Peak density during rainy season except *Amynthas diffringens* , which peaked in winter	Bhadauria and Ramakrishnan (1989)
Pasture Kumaon Himalaya	Three species: *Amynthasdiffringens, Amynthas alexandri, Eisenia fetida*	139	25	Peak density at the end of rainy season	Kaushal and Bisht (1994)
Cultivated soil Kumaon Himalaya	One species Amynthas alexandri	58		Peak density during rainy season	Kaushal *et al.* (1995)

Land use/ ecosystems	Diversity	Density (number of individuals m^{-2})	Average/ range of biomass (wet weight, g)	Trends	References
A village landscape in Garhwal Himalaya	Eight species: *Amynthas alexandri*, *Bimastos parvus*, *Metaphire anomala*, *Metaphire birmanica*, *Drawida nepalensis*, *Perionyx excavatus*, *Lennogaster pusilla*, *Octochaetona beatrix*	Traditional agro forestry system: 147; Traditional pure crop system: 132; Moderately degraded natural forest: 63; Rehabilitated Agricultural land : 63; Abandoned agricultural land: 27; Rehabilitated forest land: 8; Highly degraded forest land: 5	Traditional agro forestry system: 266; Traditional pure crop system: 199; Moderately degraded natural forest: 51; Rehabilitated Agricultural land : 21; Abandoned agricultural land: 16; Rehabilitated forest land: 24; Highly degraded forest land: 11	All species except *B. parvus* showed a strong effect of season on population size, with the highest abundance and biomass values observed during rainy season. Only *D. nepalensis* and *M. birmanica* were a little bit abundant during dry season in traditional agro forestry system.	Bhadauria *et al.* (2012)
Village landscape in Kumaon Himalaya	Eight Species: *Bimastus parvus*, *Octolasion tyrtaeum*, *Octochaetona beatrix*, *Amynthas corticis*, *Eutyophoeus festivus*, *E. nanianus*, *E. waltonii*, *Drawida* sp.	Climax forest: 526, Mixed forest: 309, 5-year-old pine: 287, 40-year-old pine: 940, grassland :353		Peak density during rainy season	Bhadauria and Ramakrishnan (2000)

Land use/ ecosystems	Diversity	Density (number of individuals m^{-2})	Average/ range of biomass (wet weight, g)	Trends	References
Village landscape in Garhwal Himalaya	Seven Species: *Amynthas corticis*, *Drawida nepalensis*, *Allobophora parva*, *Eutyphoeus pharpingianus*, *Octochaetona beatrix*, *Perionyx* sp., *Lennogaster pusillus*	108-247 in forests, 89-235 in agro-ecosystems		Abundance in pine forest- higher but diversity lower compared to oak forest	Sinha *et al.* (2003)
Nandadevi Biosphere Reserve in Garhwal Himalaya	Eight species: *Lennogaster pusillus*, *Metaphire houlleti*, *M. anomala*, *Ocnerodrilus occidentalis*, *Dendrodrilus rubidus*, *Aporrectodea calliginosa*, *Amynthas corticis*, *Drawida nepalensis*	Lower elevation: Home garden- 100, pine forest:5-10, oak forest: 5, Irrigated agriculture-5, Rainfed agriculture: 5-25 Higher elevation Home garden:10-150, Alpine pasture:5-25, Alpine forest: 10-15	9- 100	Peak density at the end of rainy season	Maikhuri *et al.* (2008)
Rubber plantations in Tripura in the north-eastern Himalaya	20 species:Glossoscolecidae- *Pontoscolex corethrurus*,Octochaetidae- *Eytyphoeus gigas*,*E.gammiei*,*E. comillahnus*,*E assamaensis*,*E. festivas*,*Eutyphoeus spp.*,*Dichogaster bolaui*,*D.affinis*,*Lenogaster chittagongnensis*, *Octochaetona beatrix*, *Megascolecidae-Metaphire houlleti*, *Perionyx* spp.,*Kanchuria sumerianus*, Kanchuria sp 1 & sp.2,Moniligastridae-*Drawida nepalensis*,*Drawida* sp-1,*Drawida* sp-2,*Gordiodrilus elegans*	10-262 (mean ~109) *Pontoscolex corethrurus-dominant* *72% of total density*,Kanchuria spp-1-*12%*, *Metaphire houlleti-3.6%*	9-100 (mean 43.4) *Pontoscolex corethrurus*-61.5% Of total biomass		Chaudhuri *et al.* (2008)

Land use/ ecosystems	Diversity	Density (number of individuals m^{-2})	Average/ range of biomass (wet weight, g)	Trends	References
Nilagiri, Biosphere Reserve: , Kerala area in the Western Ghats	Fourteen species: *Dichogaster affinis*, *Drawida modesta*, *D.barwelli impertusa*, *D.ghatensis*, *D. grandis*, *Glyphidrilus annandalei*, *Haplochetalla* sp., *Lampito mauritii*, *Megascolex insignis*, *M. triangularis*, *Octochaetona beatrix*, *Parryodrilus lavellei*, *Plutellus variabilis*, *Pontoscolex corethrurus*	Natural forests: 159; Forest plantations: 350; Palms: 258 Croplands: 640; Fallows: 124			Chandrashekara *et al.* (2008)
Nilgiri Biosphere reserve: Karnataka area in the Western Ghats	13 species: *Drawida somavarapatna*, *Drawida modesta*, *Drawida fakir*, *Drawida pellucida*, *Lampito mauritii*, *Metaphire holoutii*, *Megascolex curgensis*, *Megascolex felisiceta*, *Amynthas corticis*, *Octochaetoides castellanes*, *Pontoscolex corethrurus*, *Dendrodrilus rubidus*, *Cargia narayani*	Natural forests: 477 Forest plantations: 303 Cardamom plantations: 443 Croplands: 595 Fallows: 343			Balakrishna *et al.* (2008)

Land use/ ecosystems	Diversity	Density (number of individuals m^{-2})	Average/ range of biomass (wet weight, g)	Trends	References
A segment of Western Ghats		14-825		Earthworms mainly exhibited temporal variability while some other soil organisms were affected more by management practices	Rossi and Blanchart (2005)

to both season and land use management, with significant interactions between species, season and management practices. Flooded paddy systems had the lowest and the home gardens the highest species diversity as well as abundance (Maikhuri *et al.,* 2005). Earthworm abundance in the village landscapes at lower elevations was higher (89-940 individuals per m^2) compared to that at higher elevations (5-150 per m^2) (Table 2). In western/central Himalayan region and Tripura in north-eastern Himalaya, endogeic and endogeic-anecic dominate. In rubber plantations in Tripura established after 1962, 15 species are endogeic and only five species are epi-anecic. Further, rubber plantations gave way to exotics like *Pontoscolex corethrurus* (Chaudhuri *et al.,* 2008, 2009).

In Nilgiri Biosphere Reserve, 14 species of earthworms were recorded, with 2-8 species occurring in different land use types and total earthworm abundance in the range of 124-560 per m^2. Occurrence of endogeic *Parryodrilus lavelee* and *Pontoscolex corethrurus* in almost all land uses including degraded lands suggests that these species may have a potential for rapid restoration of soil fertility in degraded lands (Chandrashekara *et al.,* 2008). In most studies, population size has been estimated in terms of numerical abundance and not in terms of biomass.

Table 3 gives ecological category, feeding habit, and habitat and size relationship of eight species, out of 20 species of earthworms of rubber plantations raised in undulating areas in Tripura (Chaudhuri *et al.,* 2008, Chaudhuri and Bhattacharjee, 2009). This type of study will be useful for Himalayan and other regions to identify the suitable species for land use management. Functional attributes and ecological strategies of different earthworm species need to be worked out to optimize the contributions of earthworms to ecosystem services and resilience of agroecosytems.

Functional attributes: Bioturbation activity

Bioturbation refers to the biological reworking of soil and sediments, and its importance was first highlighted by Charles Darwin (1881). Bioturbation is now recognized as an archetypal example of 'ecosystem engineering', modifying geochemical gradients, redistributing food resources and microbes in soil column. Bioturbation played a key role in the evolution of metazoan life at the end of the Precambrian Era (Muys *et al.,* 2003). Earthworm casts contain more

Table 3: Ecological category, feeding habit, and habitat and size relationship of eight species of earthworms of rubber plantation in Tripura (Chaudhuri *et al.*, 2008; Chaudhuri and Bhattacharjee, 2009)

Species	Family	Size (mm) L, length; B, breadth	Color	Feeding Habit	Distribution pattern	Ecological category	Soil Temp (°c) (pH:4.4-5.2)	Soil Moisture (g %)	Soil pH	Soil Organic Matter (g %)
Pontoscolex corethrurus	Glossoscolecidae	L=72-100 B=4-5	Lightly pigmented	Geophagous	Exotic peregrine	Topsoil endogeic	19-32	10-29	4.5-4.8	1.7-2.4
Drawida Papillifer papillifer	Moniligastridae	L=45-90 B=3-4	Deeply pigmented	Phytogeophagous	Endemic	Epianecic	22-32	10-29	4.5-4.7	1.6-2.4
Drawida assamensis	Moniligastridae	L=60-80 B=4-5	Lightly pigmented	Phytogeophagous	Endemic	Topsoil endogeic	21-31	10-29	4.4-5.2	1.5-2.0
Metaphire houlleti	Megascolecidae	L=100-160 B=3-6	Deeply pigmented dorsally	Phytogeophagous	Endemic peregrine	Epianecic	21-28	14-20	4.4-5.2	1.6-2.0
Eutyphoeus comillahnus	Octochaetidae	L=70-135 B=2-4	Lightly pigmented	Geophagous	Endemic	Mesohumic endogeic	21-27	17-18	4.5-4.6	1.7-2.0
Dichogaster affinis	Octochaetidae	L=35-42 B=1-2	Moderately pigmented	Phytophagous	Exotic peregrine	Epigeic	21-28	17-21	4.8-5.2	1.8-2.0
Octochaetona beatrix	Octochaetidae	L=60-120 B=4-5	Lightly pigmented	Geophagous	Endemic peregrine	Subsoil endogeic	24-28	11..5-14	4.7-4.8	1.6-2.0
Lennogaster chittagongensis	Octochaetidae	L=40-50 B=1.5-2.5	Lightly pigmented	Phytogeophagous	Endemic peregrine	Topsoil endogeic	24-30	16-18	4.7-4.8	1.5-2.0

water soluble aggregates and higher nutrient concentrations than the surrounding soils. Soils with earthworms drain 4 to 10 times faster than soils without earthworms (Nijhawan and Kanwar, 1952; Guild, 1952, 1955; Low, 1955; Kale and Krishnamoorthy, 1980; Dash and Patra, 1979; Petersen and Luxton, 1982; Bhadauria and Ramakrishnan, 1989, 1996).

Bhadauria and Ramakrishnan (1991) estimated cast production at a rate of 20 tons, 35 tons, 40 tons per hectare per year in a 5-year-old pine forest, a 35-year-old pine forest, and a sacred grove (close to climax vegetation), respectively, in the north-eastern hill region of India. Chaudhuri et al. (2008) estimated cast production at a rate of 2.51 ton per ha per year in rubber plantations in Tripura in the north-eastern India. These rates of cast production are substantially lower than the rates of 77-141 tons hectare per year reported in temperate/ tropical ecosystems (Roy, 1957; Satchell, 1967; Dash and Patra 1979; Reddy, 1983). Estimation of cast production in different land uses would indicate the functional role of earthworms but such studies covering different land-uses and eco-regions in India are lacking.

Land use change and earthworms

Bhadauria and Ramakrishnan (2005) worked on earthworm community structure in relation to land use change in shifting agricultural landscapes in the north-eastern Himalaya (Fig. 2), Chaudhuri *et al.* (2008, 2009) in rubber plantations in Tripura, Chandrasekara *et al.* (2008) in Nilgiri Biosphere, Senapati *et al.* (1994) in tea gardens in south Indi and Mishra and Dash (1984) (Table 4), Dash and Senapati (1991) ; Behera *et al.* (1999) and Senapati *et al.* (2005) in Orissa in south-east India.

Figure 2: Land use change and earthworm community structure in north-eastern Himalaya (Bhadauria and Ramakrishnan, 2005)

Table 4: Earthworm species richness in Easternghat land uses (Based on Mishra and Dash, 1984; Senapati *et al.,* 2005)

Sl. No.	Species	Land Uses				
		Shifting Cultivation				
		Natural Forest	Disturbed fallow	8-year old phase	Cropping plantation	Eucalyptus
1.	Drawida calebi	+	+	-	-	-
2.	Drawida willsi	+	-	-	-	-
3.	Eutyphoeus incommodes	+	-	+	+	-
4.	E.waltoni	-	-	-	+	-
5.	Eutyphoeus sps.	+	-	-	+	-
6.	Lampito mauritii	+	+	-	-	+
7.	Lennogaster dashi	-	-	-	-	+
8.	Lennogaster pusillus	+	-	+	+	-
9.	Ocnerodrilus occidentalis	+	+	-	-	-
10.	Octochaetona surcnsis	+	+	-	-	-
11.	Pellogaster bengalensis	+	+	+	+	-
12.	Ramiella bishambari	+	+	+	-	-

Based on these studies, some generalizations can be made. Conversion of natural forests to shifting agriculture and plantations results in some loss of earthworm species richness together with changes in composition of soil fauna community structure and function. Exotic and native species coexist in natural and derived (managed) ecosystems but exotics are less frequent in primary forests. Endopolyhumics in primary forests and Endomesohumics in derived ecosystems (man-managed) dominate the earthworm community structure in north-east India. However in central Himalayas, endomesohumic earthworms dominated irrespective of land use changes from primary forest to grassland and fallows. Land use change from forest to agro ecosystem favoured endomesohumic possibly due to high input of farmyard manure in settled agroecosystems in central/ western Himalaya.

Senapati *et al.* (2005) observed proliferation of termite populations following land use intensification in tea plantations and Senapati *et al.* (1994) and Senapati (1997) suggested that termite-earthworm biomass ratio can be used as a sensitive index to land use change. This hypothesis is yet to be tested in wide variety of ecosystems.

Conclusions

Conversion of natural forest results in decline of earthworm species richness and abundance. Organic inputs in the form of manure and crop residues in the derived systems help in restoring earthworm fauna. Termites may proliferate at the expense of earthworms in these derived systems. Periodically inoculation of endogeic and endo-anecic earthworms is likely to restore soil quality in degraded ecosystems. The selection of earthworm species should centre on the naturalized species in a particular site irrespective of whether the species is native or exotic.

Acknowledgements

This publication arises from our association with and support from Tropical Soil Biology and Fertility Insitute of CIAT (TSBF-CIAT), TSBF-UNEP-GEF programme on belowground biodiversity and MMSEA-SLM network of United Nations University Institute of Sustainability and Peace (UNU-ISP), Tokyo.

References

Behera, B. Giri, S., Dash, N. C., Sahu, J. and Senapati, B. K. 1999. Earthworm bioindication of forest land use pattern in western Orissa. *Indian Forester.* 125: 272-281.

Bhadauria, T. and Ramakrishnan, P. S. 1989. Earthworm population dynamics and contribution to nutrient cycling during cropping and fallow phases of shifting agriculture (Jhum) in northeast India. *J. Applied Ecology*. 26: 505-520.

Bhadauria and Ramakrishnan, P. S.1991. Population dynamics of earthworms and their activities in forest ecosystems of north-east India. *J. Tropical Ecology.* 7: 305-318.

Bhadauria, T., Ramakrishnan, P. S. and Srivastava, K. N. 1997. Population dynamics of earthworms during crop rotation under rain fed agriculture in central Himalayas, India. *Applied Soil Ecology.* 6: 205-215.

Bhadauria, T. Ramakrishnan, P. S. and Srivastava, K. N. 2000. Diversity and Distribution of endemic and exotic earthworms in natural and regenerating ecosystems in the central Himalaya, India. *Soil Biology and Biochemistry.* 32: 2045-2054.

Bhadauria, T. and Ramakrishnan, P. S. 2005. Impact of land use-land cover change on earthworm community structure in India. Pages 57-65.In: Ramakrishnan, P. S., Saxena, K. G., Swift, M. I., Rao, K. S. and Maikhuri, R. K. (Editors) Soil Biodiversity, Ecological processes and Landscape Management. Oxford and IBH Publishers, New Delhi, India.

Bhadauria, T., Kumar, P., Kumar, R., Maikhuri, R. K., Rao, K. S. and Saxena, K. G. 2012. Earthworm populations in a traditional village landscape in Central Himalaya, India. *Applied Soil Ecology.* 53: 83-93.

Chandrasekara, U. M., Balasundaram, M., Sankaran, Sujatha, M. P., Varma, R. V., Senapati, B. K. and Sehgal, M. 2008. Conservation and Sustainable Management of Below-ground Biodiversity in the Kerala part of Nilgiri Biosphere Reserve-phase I, Final Technical Report, Kerala Forest Research Institute, Thissur, India. p.119.

Chaudhuri, P. S., Nath, S. and Paliwal, R. 2008. Earthworm population of rubber plantations (*Hevea brasiliensis*) in Tripura India. *Tropical Ecology.* 49: 225-234

Chaudhuri, P. S. and Bhattacharjee, S. 2009. Reproductive biology (cocoon structure, incubation period, emergence pattern and fecundity) of eight tropical earthworm species of rubber plantations in laboratory cultures. *J. Tropical Ecology* (In press)

Choudhuri, P. S. 2009. Personal Information

Darwin, C. 1981. The formation of vegetable mould through the action of worms with observations on their habits. John Murray, London

Dash, M. C. 2012. Charles Darwin's Plough. Tools for Vermitechnology. I.K. International Publishing House Pvt. Ltd. New Delhi. p. 185.

Dash, M.C. and Dash, S. P. 2008. Conservation and sustainable management of belowground biodiversity: A review on the functional role of soil fauna in Indian ecosystems with particular reference to earthworms. *International J. Ecology and Environmental Science.* 34 (3) :223-243.

Dash, M. C., Saxena, K. G. and Giri, S. 2009. Vermitechnology for Wasteland Reclamation, Plant Productivity and Composting: A Review in Indian Context. *International J. Ecology and Environmental Sciences.* 35 (2-3):163-185.

Dash, M. C. and Patra, U. C. 1979. Worm cast production and nitrogen contribution to soil by a tropical earthworm population from a grassland site in Orissa, India. *Review of Ecology and Biology of Soil.* 16: 79-87.

Dash, M. C. 1978. The role of earthworms in the decomposer system, Pages 399-409, In: Singh, J. S. and Gopal, B. (Eds) Glimpses of Ecology, International Scientific Publishers, Jaipur, India

Dash, M. C. and Patra, U. C. 1977. Density, biomass and energy budget of a tropical earthworm population from Orissa, India. *Review of Ecology and Biology of Soil.* 14: 461-471.

Dash, M. C. and Senapati, B. K.1985. Vermitechnology potentiality of Indian earthworms for vermicomposting and vermifeed.Pages 61-69.In: Mishra, H.K. and Kapur, K. (Editors) Proceedings of National Seminar on Current Trends in Soil Biology. Haryana Agriculture University, Hissar, India.

Dash, M. C. and Senapati, B. K. 1991. Impact of Eucalyptus litter on earthworm activity. In: Advances in Management and Conservation of Soil Fauna. Veeresh, G.K., Rajgopal, D. and Viraktamaath, C.A. (Editors), Oxford and IBH Publication Company Private Limited, New Delhi. pp. 349-355.

Guild, W. J. Mc.L 1952. Variations in earthworm numbers within field population. *Journal of Animal Ecology.* 21:169

Guild, W.J.Mc.L.1955. Earthworms and Soil Structure. Pages 83-98, In: Kevan, D.K.Mc.E. (Editor) Soil Zoology, Butterworth's, London.

Julka, J. M. and Paliwal, R. 2005a. Diversity and Biogeography of Indian Earthworms, In:Proceedings of National Level Workshop on Vermitechnology Transfer to NSS Programme Officers, Department of Biochemistry and Zoology, Kongunadu Arts and Science College, Coimbatore, India, 5-21.

Julka, J. M. and Paliwal, R. 2005b. Distirbution of earthworms in different agro-climatic regions of India. In: Soil Biodiversity, Ecological Processes and Landscape Management. Edited by P.S. Ramakrishnan, K.G. Saxena, M.J. Swift, K.S. Rao and R.K. Maikhuri. Oxford and IBH Publishing Company Private Limited, New Delhi. 3-13

Julka, J. M. and Mukheree, R. N. 1984. Mass migration of earthworm in the Simla hill. Page 347-351, In: Proceedings of workshop on High Altitude Entomology and Wildlife, Zoological Survey of India, Calcutta.

Julka, J. M. 2008. Personal communication.

Kale, R. D. and Dinesh, M. S. 2005. Earthworm: Components of Soil Biota, In: Proceedings of National Workshop, Kerala Forest Research Institute, Peechi, Kerala, India, pp. 313-330.

Kaushal, B. R., Bisht, S. P. S. and Kalia, S. 1995. Population dynamics of earthworm *Amynthas alexandri* (Megascolecidae: Annelida) in cultivated soils of the Kumaon Himalayas. *Applied Soil Biology.* 2: 125-130.

Low, A. J. 1955. Improvement in the structure of soils under leys. *Journal of Soil Science.* 6: 179-199.

Maikhuri, R. K., Rawat, L. S., Negi, G. S., Dangwal, D., Chandra, A. and Rao, K. S. 2005. Macro fauna in Nanda Devi Biosphere Reserve. In: The

Proceedings of the National Workshop on Conservation and Sustainable Management of Belowground Biodiversity, Kerala Forest Research Insitute, Peechi, India, pp. 422-431 (unpublished document).

Mishra, P. C. and Dash, M. C. 1984. Population dynamics and respiratory metabolism of earthworm population in subtropical woodland of western Orissa, India. *Tropical Ecology.* 25:103-116.

Muys, B., Beckers, G., Nachtergale, L., Lust, N., Merck, R. and Granval, P. 2003. Medium term evaluation of a forest soil restoration trial combining tree species change, fertilization, and earthworm introduction. *Pedobiologia.* 47: 772-783.

Mishra, K. C. and Ramakrishnan, P. S. 1988. Earthworm population dynamics in different Jhum fallows developed after slash and burn agriculture in north-eastern India. *Proceedings Indian Academy of Sciences (Animal Sciences).* 97: 309-318.

Myers, N., Mittermeier, R. A., Mittermeier, C. G., da Fonseca, G. A. B. and Kent, J. 2000. Biodiversity hotspots for conservation priorities. *Nature.* 403: 853-858.

Nijhawan, S. D. and Kanwar, J. S.1952. Physicochemical properties of earthworm castings and their effect on productivity of the soil. *Indian Journal of Agricultural Sciences.* 22: 357-373

Petersen and Luxton, M. 1982. A Comparative analysis of soil fauna populations and their role in decomposition process. *Oikos.* 39: 287-388.

Ramakrishnan, P. S. Saxena, K. G., Swift, M. J., Rao, K. S. and Maikhuri, R. K. 2005. Soil Biodiversity, Ecological Processes and Landscape Management. Oxford and IBH Publishing Company Private Limited, New Delhi. p.302.

Reddy, M. V. 1987. Seasonal structure of earthworm (*Pheretima* sp.) Populations in humid tropical deciduous woodland. In: Bonvicini Paglial, M and Omodeo, P (Editors) On Earthworms, Selected Symposia and Monographs, U. Z. I., Mucchi, Modena. pp. 191-197

Reddy, M. V. and Alfred, J. R. B. 1978. Some observations on the earthworm population and their biomass in a sub-tropical pine forest soil. In: Edwards, C. A. and Veeresh, G. K. (Editors) Soil Biology and Ecology in India. University of Agricultural Sciences, Bangalore. pp. 78-82

Satchell, J. E. 1967. Lumbicidae. In: Burges, A and Raw, F.B. (Editors), Soil Biology, Acadfemic Press, London. pp. 259-322

Senapati, B. K. and Dash, M. C.1981. Effect of grazing on the elements of production in vegetation and Oligochaeta of a tropical pasture, Review of Ecology and Biology of Soil. 18: 487-505.

Senapati, B. K., Panigrahi, P. K. and Lavelle, P. 1994. Macrofaunal status and restoration strategy in degraded soil under intensive tea cultivation in India. In: Transactions of 15th International Soil Science Congress. *Acapulco and Mexico.* 4 (a) : 64-75

Senapati, B. K.1997. Innovations in Soil management through earthworm technology: A case study in tea agroecosystems, In: Modern Approaches and Innovations in Soil Management. Bagyaraj, D. J., Verma, A., Khanna, K. K. and Kheri, H.K. (Editors), Rastogi Publications, Meerut, pp. 282-300.

Senapati, B. K., Lavelle, P., Panigrahi, P. K., Giri, S. and Brown, G. G. 2002. Restoring soil fertility and enhancing productivity in Indian tea plantations with earthworm and organic fertilizers. International Technical Workshop on Biological Management of Soil Ecosystems for Sustainable Agriculture. *Brazilian Agriculture Research Corporation, Brazil.* 182: 172-190.

Senapati, B. K., Julka, J. M. and Lavelle, P. 2005. Impacts of Land use-land cover change on soil fauna in south and south-east India. In: Ramakrishnan, P. S., Saxena, K. G., Swift, M. J., Rao, K. S. and Maikhuri, R. K. (Editors), Soil Biodiversity, Ecological Processes and Landscape Management, Oxford & IBH Publishing Co. Pvt. Ltd., New Delhi. pp. 67-75

Sinha, B., Bhadauria, T., Ramakrishnan, P. S., Saxena, K. G. and Maikhuri, R. K. 2003. Impact of landscape modification on earthworm diversity and abundance in the Hariyali sacred landscape, Garhwal Himalaya, *Pedobiologia.* 47:357-370.

State of Forest Report, Government of India, 2005 and 2009.

7

Bioresources as a Tool for Food Security and Sustainable Development for Rural Livelihood in India in The Context of Industrial Development and Environmental Protection: An Overview

Introduction

Max Muller, the celebrated German Scholar once wrote 'if we were to look over the whole world to find out the country most richly endowed with all the wealth, power and beauty which nature can bestow - in some parts a veritable paradise on earth - I should point to India'. What he wrote of India, in fact, is truer for the Rural India which is a unique depository of all those beautiful things that God and human hands created for the posterity to draw inspiration for a better tomorrow. India is a country of villages. As per 2001 census, on the average 72.22% people lives in villages. In some states more than 85% people live in villages (Himachal Pradesh, 90.21%, Bihar, 89.53%,

Sikkim, 88.9%, Assam, 87.28% and Orissa, 85.03%). The livelihood options and rural economy in Indian villages are largely bio-resource based, especially on agriculture, home garden, fishery, goatery, piggery, poultry, diary and artisan work based on bamboo, cane, wood and other bio-resources. In view of this, conservation of traditional ecological knowledge and management of bio-resource is most important.

This paper reviews the bio-resource and livelihood options in rural India. Since the management of bio-resources are linked with land, soil and water, these aspects are also reviewed in this paper. Recent thrust on industrialisation in resource rich states like Jharkhand, Orissa, Chhatisgarh, Rajsthan, Uttarakhand and some other states threaten the traditional livelihood options and therefore need implementation of EIA and EMP protocol very strictly and judiciously.

Resource classification

Resources are basically of two kinds, (a) renewable, and (b) non-renewable. Some resources, such as plants (crops, forests, medicinal plants etc.) and animals (milk and meat producing) are renewed from time to time because they have a life-cycle, and a continuous harvest is possible provided these resources are well managed. These resources sustain life. Bio-resources become extinct if not managed properly. These resources require land, soil and water for their growth and continuance. (Dash and Dash, 2009)

Land resources

India with a land area of 3, 28, 8000 km^2 (about 2.4% of the world), supports 16% of the world's population. Land is now under great pressure due to the huge increase in population. There were about 238 million people in India in 1901 and have grown five times in 100 years (about 1200 million). With the present rate of population growth the per capita land resource in India may be reduced to <0.25 hectare in very near future.

About 44% of our land is used for agriculture, 23% is covered with forests, and 4% is used for pasture and grazing fields, 8% for housing, agro forestry, and industrial areas, and roads and so on. 14%of land is barren and about 7% is used for miscellaneous purposes. The rapid increase of urbanization and migration of population from rural area to towns and cities has led to conversion of agricultural land for housing, construction of office buildings, industries, and so forth. The rational use of land resources is possible only by adopting an integrated land-use policy through scientific survey of all land resource taking the village as a unit and apportioning land for both short and long-term requirements for agriculture, forestry, grazing, water bodies and fisheries, human settlements, roads, industries, and so on. The state governments play important role. There should be legislative control of land use.

The fourteen physiographic zones of India are listed in Fig. 1 and below:

1. Western Himalayas (WH), 2.EasternHimalayas (EH), 3. North East (NE), 4.Northern Plains (NP), 5.Eastern Plains (EP), 6. Western Plains, 7.Central Highlands (CH), 8.North Deccan (ND), 9.South Deccan (SD), 11 Western Ghats (WG), 12.Eastern Ghats (EG), 13.West Coast (WC), 14.East Coast (EC).

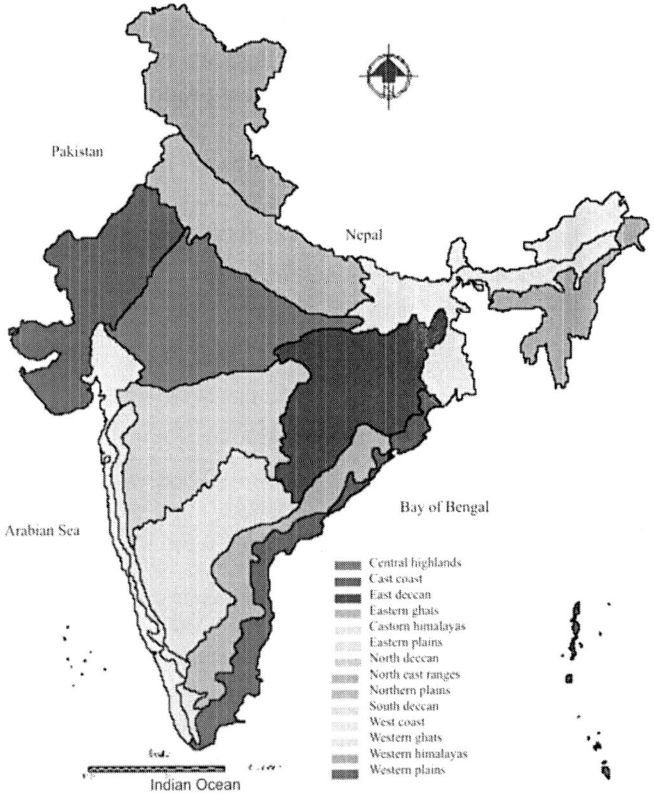

Figure 1: The physiographic zones of India (State of Forest Report, 2005)

These physiogeographical zones provide the livelihood support to large majority of the population. All the Bio-resources are located in these zones. However each zone is unique with regard to bio-resource and availability of water and other resources.

Forest resources

The forest cover is important bio-resource for the people, especially for the village people and ecosystem people. Forest cover includes all lands which have a tree canopy density of 10 percent and above with area 1 ha or more. The minimum mapped area of 1 ha of the forest cover corresponds to the cartographic limit (a polygon of the size 2 mm x 2 mm) on a map at 1:50, 000 scale. This definition is based on the scale of interpretation (1:50, 000), the optimum size of cluster of pixels, resolution of digital satellite data used for the mapping (pixel size 23.5m x 23.5m) and the technique employed for image interpretation (State Forest Report, 2005-Government of India). The forest cover mapping presented in the 'State Forest Report' does not make any distinction between the origin of tree crops (natural or man-made) or tree species. Also, it does not recognize the type of land ownership or land use and legal status of land under the forest cover. Thus, all species of trees (including bamboos, fruits, coconut, palms, etc.) and all types of lands (forest, private, community or institutional) satisfying the criteria of canopy density of more than 10 percent have been delineated as forest cover while interpreting the satellite data. The basis of classifying forest cover, the ministry of Environment and Forests follow the principle given in Table 1.

Table 1: The Basis of Classification of Forest Cover

Very Dense	
Forest	All lands having tree cover with canopy density >70%
Moderately Dense Forest	All lands having tree cover with canopy density between 40% and 70%
Open Forest	All lands having tree cover with canopy density between 10% and 40%
Scrub	All lands with mainly small and stunted trees with canopy density <10%
Nonforest	Any area not included in the above classes

Forest Cover: 2005 Assessment

The total forest cover of the country as per 2005 assessment is 677, 088 km² and this constitutes 20.60 percent of the geographic area of the country. Of this, 54, 569 km² (1.66 %) is very dense forest, 332, 647 km² (10.12 %) is moderately dense forest, while 289, 872 km² (8.82 %) is open forest cover. The scrub accounts for 38, 475 km² (1.17 %).

While computing the percentage of forest cover of the country, the total geographical area of 3, 287, 263 km² is taken. A closer analysis of this reveals that a sizeable part of the country's area lies in high altitude mountainous region under permanent snow/glaciers, steep slopes and rocks which are not available for tree planting due to climatic and physical reasons. As per a recent study of FSI, about 1, 83, 135 km² areas in the States of Arunachal Pradesh, Himachal Pradesh, Jammu Kashmir, Sikkim and Uttarakhand have been found above 4, 000 m altitude where climatic and edaphic conditions limit tree growth. If this part of the geographical area of the country is excluded for the purpose of forest cover analysis, the forest covers of the country in terms of percentage to the geographical area is much less. Forest cover in hills is essential to maintain ecological balance and environmental stability as it prevents soil erosion and land degradation. The State/ UT wise forest cover in the country shows that Madhya Pradesh with 76, 013 km² has the largest area under forest cover, followed by Arunachal Pradesh (67, 777 km²), Chhattisgarh (55, 863 km²), Orissa (48, 374 km²) and Maharashtra (47, 476 km²). Considering the proportion of geographic area under forest cover, Mizoram has the maximum percentage of 88.63%. The recent forest policy of Indian Government envisages 33% of land under forest cover. This will also

Table 2: Energy Sources for Rural India

Source	% of the total energy consumed
Renewable	
Wood	68.5
Animal dung	08.3
Others	03.4
Non-renewable	
Oil products	16.9
Coal	02.3

Table 3: Production and productivity of some crops in India (Source : Prasad, 2009)

Crop	Year								
	1967-1968 (Year of green revolution						2005-2006 % increase over 1967-68		
	Area Million ha		Producion million tons		Yield ton/ha		Area	Production	Yield
Wheat	15 (26.4)		16.5	(69.3)	1.1	(2.6)	76.7	144	110
Rice	36.4	(43.7)	37.6	(91.8)	1.0 (husked clean rice		20.0	166	136
Maize	5.6 (7.6)		6.3 (14.7)		1.1	(1.9)	35.7	133	73
Pulses	22.6	(22.4)	12.1	(13.4)	0.5	(0.6)	-	11	20
Total Food grains	121.4 (120.0)		95.0	(208.3)	0.8	(1.6)	-0.01	100	119
Major Nine Oilseeds	15.7	(27.9)	8.3 (28.0)		0.5	(1.0)	77.7	237	100

increase the sink area for green house gases and will create avenues for other benefits. Massive effort is required to achieve this and to increase the cover from present about 20.60% to 33%. 91, 663 km2, which constitutes 2.79 percent of the country's geographic area. The estimates of tree cover for each physiographic zone is given in the map. It is observed that the tree cover is the maximum in East Deccan (11, 293 km2), followed by Northern Plains (10, 747 km2) and West Coast (8, 307 km2). West Coast has maximum percentage of tree cover (6.85%) with respect to geographical area followed by Western Ghats (5.37%) and East Coast (4.84%). Eastern Himalayas has the lowest tree cover of 255 km2, although the area is full of forests. There is need of increasing the tree cover in Himalayas, especially in Eastern Himalayas and in the urban centres of the country.

Soil resources

Tree cover

The total tree cover of the country has been estimated to be

Table 4: Fish production in India (tons per annum)

	Fish Production (t)	Capture (t)	Aquaculture (t)
1997	5 485 270	3 620 948	1 864 322
1998	5 381 977	3 473 492	1 908 485
1999	5 686 963	3 552 149	2 134 814
2000	5 668 631	3 726 427	1 942 204
2001	5 936 931	3 817 092	2 119 839
2002	5 932 542	3 745 353	2 187 189
2003	6 033 870	3 720 899	2 312 971
2004	6 190 314	3 391 010	2 799 304
2005	6 658 010	3 691 364	2 966 646
2006	7 014 143	3 844 838	3 169 305
2007	7 308 233	3 953 476	3 354 757

(Source FAO, 2009)

Soil supports the life system and is formed through the combined action of parent rock material, climate, weathering process, vegetation and the decomposition process. Soil formation is an extremely slow process, taking thousands of years. It is important in that soil stores water and nutrients for biota including plants. This resource is now

Table 5: Livelihood support in Orissan villages (Dash, 2007, 2009)

Social category	Bioresource Used	Issues
Landless and Wage earner: predominately SC and STs, financial weaker section (OBCs etc)	Forest, and forest products, Domesticated goat, poultry,Rivers,Nallas for fish and depend on upper-class/government/Private project work for wages	Food security, lack of provision for health issues, largely depend upon community common- property bioresource
Marginal and small farmers(Financially weaker section)(usually with < 1acre of land)	Can not survive on land alone, non-monsoon time-depend on common-property bioresource,domesticated goat, cowand poultry	Unable to harness benefits because of small holdings, and lack of adequate water resource/irrigation and above cited issues
Ecosystem Dependent people (Forest dependent communities, Fishing communities)(Tribals,Fishermen)	Fully bioresource based livelihood, Home garden, Animal rearing etc.	Seasonal unemployment, Exposure to risks, Income variable, Accident/health risk- can not afford insurance
Displaced people due to industrialisation	Bioresource,Employment, Entrepreneurship etc	Loss of land and livelihood, unable to adjust to new livelihood options,and new homes etc.
Widows, Destitute, Persons with disability, Disaster affected	Bioresource(Community common-property based),Domestic help,	Unable to cope with problems, Food security, Health security, Identity crisis
Artisans	Bioresource (Bamboo,Cane,Wood etc)	Finance, Marketing etc
Miscellaneous (small %)	Employment, Small Business,	Health issues and other problems

Table 6: Environmental Liability of Livelihood options in villages

Environmental Stress	Criteria	Remark	Pollution
Low	Bio-resource based livelihood,	Materials are biodegradable,(small tea stalls,tailoring,etc),	Foul smell,
	Non-biodegradable substances are small	Cooking by biomass based fuels in quantity and reusable.	Air pollution from biomass-cooking
Medium	Bio-degradable substances with high organic and salt loading are produced,	Fish drying activity, Bakery and Pitha making,	Foul smell,
	Non-biodegradable and non-hazardous substances are produced in small quantity and recycling is possible to a large extent, fossil fuels in small quantity are used.	BOD of the effluent not exceeding litre, Workers involved in Agricultural 100mg/ activities using tractors, diesel irrigation pumps seasonally,	Air pollution by exhaust gases and Suspended particulate matter,
		stone crushers etc	Health risks
High	Non-biodegradable, hazardous and toxic substances that create severe and adverse environmental impacts,	Chemical fertilizer and pesticides used in agricultural field, Large brick kilns,	Workers exposed to severe pollution,
	Natural resources are used in significant quantities which are not sustainable, Fossil fuel is routinely used	Bio-resource based furniture making, Furnace and boilers using coal	Depletion of resource due to unsustainable practices

threatened because of large scale deforestation and erosion of the topsoil by wind action, floods and water flow. Soil erosion is a big problem in most of the states, especially in Himalayan region.

Water resources

Of the total water resource of the earth, 97.3% is salt water and the rest fresh water. The fresh water 2.7% amounts to about 1.4 billion km^3 of water.

Of this, about 77.2% is permanently frozen, 22.4% occurs as ground water and soil moisture, 0.35% is found in lakes and wetlands, and 0.01% in rivers and streams (World Resources, Basic Books, New York 1986). The total volume of ground water found in underground reservoirs, called aquifers, is estimated to be 42.3 x 10^{10} m^3. Agriculture uses the maximum amount of water in the world and puts lot of pressure on ground water. Excessive use of ground water depletes aquifers, lowers the water table and may lead to salinisation, water logging and alkalinisation of soils. It has been estimated (Dash, 2006) that the fresh water needs (drinking and food) of a person are about 2.7L per day (1, 000L or 1m^3 per year). India receives about 3 trillion m^3 of water from rainfall, which amounts to about 105 to 117 cm annually. This is a huge water resource and the largest in the world. But almost 90% of this precipitation falls between mid-June and October. Fourteen major river systems, such as the Ganges, Narmada, Brahmaputra, Mahanadi, Cauvery, Krishna and Godavari account for 85% of surface flow and share 83% of the drainage basin. They serve 80% of the total population. There are about 100 medium and minor rivers. The storage capacity is 3.65 million m^3. Of the total annual precipitation, India utilizes only 10%, which may increase to about 26% by 2025.We need to apply improved technology to increase water harvesting and storage capacity and ground water utilization. Of the total water used in India, 92% is for irrigation and 8% for industrial and domestic use (Sharma, 1987).

The Central Ground Water Board has estimated that the available ground water in India is 210 billion m^3 and the annual utilization potential is about 42.3 million hectare, only one-fourth of which is used at present. However, the ground water availability is not adequate in provinces like Tamil Nadu and Andhra Pradesh and excessive ground water use (unsustainable use) for agriculture in Punjab, Haryana, Delhi and Rajasthan has depleted ground water resource. More than 26 cubic miles of ground water vanished from aquifers in these states due to excessive agriculture use and may bring collapse in agriculture production unless remedial measures taken on priority basis (NASA satellite data-Yahoo internet news dated 14 August 2009). Adequate ground water recharge and shifting to non water intensive crops are important to address the problem. Due to large scale deforestation and monsoon failure, there is a regular occurrence of drought in Kalahandi district and some other districts of Orissa, in Jharkhand, Bihar, Karnataka, Rajasthan, and Maharashtra.

The supply of drinking water in Indian villages is not adequate, especially in summer. There is no organized water supply in most of the villages and even in some small towns. Although there is good precipitation in India, sufficient care is to be taken to manage this water efficiently.

The wetland, coastal and marine ecosystems are highly productive and thousands of coastal villages depend on the ecosystems for their livelihood.

The coastal areas, apart from being environmentally sensitive and highly productive, are extremely attractive for human population and, therefore, for economic development.

Sixty percent of the world's population lives within 60 km of the coastlines, and this figure is forecasted to rise to 75% within the next three decades. The average width of the coastal zone on the terrestrial side is considered to be 60 km worldwide.Few thousand coastal villages exist in east coast and west coast of India. The livelihood options of these villages are predominantly marine and brackish water fishery.

It is essential to manage coastal zone, balancing and integrating demands of various sectors at the same time facilitate much needed awareness about its values and functions and promote its wise use by the local communities who depend on it for their livelihood

A variety of landform features are observed in this part such as spits, bars, lagoon, lakes, creeks, swamps, beaches, tidal flats including the classical mangrove swamps.

The coastal processes are complex and the coastal forms are very dynamic as changes happen regularly. India has about 7500 km of coast line and a number of coastal wetlands adorn the coastal zone of the country.

Mangroves forests are one of the most productive wetlands and bio-resource on earth and provide livelihood to thousands of village people.

In West Bengal, Orissa and Adaman-Nicober Islands mangrove forest and swamps are found. The mangrove forests are under threat due alteration of fresh water flow and from the anthropogenic pressure. Their protection and conservation are the need of the time (Dash and Kar, 1990; Dash, 2006)

Sea grasses constitute of about 60 species of bio- resource of underwater marine ecosystem. Fourteen species of sea grasses belonging to seven genera are so far recorded in India, out of which three genera encompassing six species are recorded from Chilika lagoon (Pattnaik, 2006). Sea grass ecosystems provide habitats for a wide variety of organisms, especially fish, and shrimps and birds.Sea grasses are very sensitive to some human impacts and are important indicators of the health of the marine environment. Sea grasses with high protein content have commercial value. Many coastal village people depend on them for livelihood.

Bio-resources

These resources are forests, crops, other plants, wildlife, domestic animals, and aquatic plants and animals, and their products yielding food, energy, biogas, and so on.

Biomass

These are, however, renewable sources of energy. Fuel wood demand in India was about 300 million tons in 2005 AD. Natural forests provide 75-100 million tons and the rest of the demand is met with fuel wood plantations, which require lot of effort and sound management strategy.

Biogas: India has >391 million livestock, 185 million cattle, 61 million buffaloes, 45 million sheep, 97 million goats, 1 million horses, 1 million camels, and 1 million other livestock. These animals produce a huge quantity of dung, which can be utilized for biogas production. Water hyacinth, Hydrilla, duck weeds and algae can act as supplements for production of biogas. Besides cooking, biogas can be utilized to produce steam, which can be used to run machines and turbines for generating electricity. Left-over slurry can be utilized as manure.

Sharma (1987) mentions that some 1, 000 to 1, 500 million tons of wet animal dung is available per year in India and at a 66% collection rate, 22, 425 million/m^3 of biogas can be produced. The slurry can produce 206 million tons of organic manure per year, equivalent to 1.4 million tons of nitrogen, 1.3 million tons of phosphate and 0.9 million tons of potash as fertilizers.

At present the amount of dung production is much more. There are about 0.61 million biogas plants in India now. Biogas generation, utilization and management require top priority to address the energy requirement problem in rural India and if implemented in large scale will help forest conservation.

Biomass Energy: Some conventional energy resource, such as wood- biomass fuels are also renewable. The principal energy sources for rural India are given in Table 2.

The firewood demand in India was about 300-350 million tons in 2, 005.Biomass such as firewood, agricultural residue, bagasse, crop stalks, coconut shells, waste from agro based industries, and animal dung can be used to produce power. Direct burning of these wastes is inefficient and leads to pollution. If these wastes are combusted in a gasifier at low oxygen and high temperature, can be converted into a gaseous fuel called producer gas. This gas, although has a lower calorific value compared to natural gas or liquefied petroleum gas, can be burned with high efficiency and without emitting smoke.

India generates > 600 million tons of agricultural residues per year. These wastes can generate 79000 MW of power if all of this waste is gasified (SPAN, 2006). It is feasible to set up a biomass based power plants of 10 to 20 MW in every village block, consisting 100 villages.

Forests as Bio-resource

These are natural ecosystems dominated by trees. About a third of the world's land surface is covered by forests, of which tropical forests constitute about 50%. Forest ecosystems are great resources, since they provide habitat for wild life, fuel wood, fodder, fiber, fruit, herbal medicines, timber, and several raw materials, to rural people in India and these are primarily used in bamboo and wood-based cottage industries, including paper and pulp industries. A climax forest may account for 400 ton of dry matter per hectare. The large animal biomass may account for about 1, 000 kg per hectare.

Dependence of the rural society on forests

Forests provide timber, firewood, nuts, fruits, and seeds, medicinal plants etc. to the villages in India. Forests shape natural environment of villages by providing ecosystem services like prevention of soil erosion, flood, etc and influencing such factors as temperature, humidity and precipitation and as a sink for green house gases and some pollutants. Forests provide suitable habitats for a number of plant and animal species. They help in maintaining a broad genetic base from which future strains and varieties could be developed.

Nearly 23% of the total land area of India (328.8 million hectare) is occupied by forests, amounting to about 74.8 million hectares. However, analysis of satellite photographs indicates that the forest cover may now only be 14% (11% closed forests and 3% degraded woods).

It has been estimated that since independence, India has lost about 45 million hectares of good forests, of which only 6 million hectares have been replaced (Sharma, 1987). The principal factors for deforestation are an increase in human population and livestock and consequent increased demand for timber, fuel wood, and grazing. Urbanization and industrialization are important factors causing the destruction of forests. Other important causes are the construction of roads and mining activities. The construction of hill roads, particularly in the Himalayan region, has destroyed many forestlands. One estimate shows that some 683, 672 hectares of land were under mining activity in 1973. Mining activities are still continuing in many states in full swing – the bauxite, coal, iron ore, chromites, lime stone, dolomite-mining in Orissa, Jharkhand, Chhatisgarh, Rajsthan and some other states is a case in point.

Wood consumption worldwide amounts to >3.2 billion m^3. Of this total some 46 and 54% respectively are consumed by industry and as firewood in developing countries. The corresponding figures for developed countries are 84 and 16%, and most of the wood being consumed as firewood. In India, fuel wood production was 213 million m^3 in 1983, about 23% higher than the figure for 1973. (FAO year book of forest products, Rome, 1985). Paper industries consume a huge amount of bamboo, which is now in short supply in peninsular India. Paper mills now depend upon supply from North-east, especially Arunachal Pradesh (Sharma, 1987).

The Himalayan mountain areas amount to 12, 49, 000 km^2 which form about 38.4% of India's land resource. Deforestation is now a major problem in the Himalayan regions. It has caused soil erosion and other damages. Besides, a number of wood-based industries have come up in this region as raw materials are easily available. The growth of industries has also caused air and water pollution. No detailed survey on the rich wealth (fauna, flora, minerals, water resource, etc.) of Himalayan ecosystems has been made. Recently WWF (2009) have reported 350 new species of biota from eastern Himalayas.Village ecosystems in general and mountain village systems in particular are under severe stress.Fast depletion of forests including mangroves by illegal felling, practice of Podu in many areas without taking remedial measures create stress among village people, especially in tribals.

Social forestry involves plantations on government, private and village panchayat land, roadsides, canal-banks, along railway lines, land not suitable for agriculture and so on. The aim is to grow fast growing trees, which are commercially valuable. These trees are meant to raise firewood for people, pulpwood for the paper industry, and fodder for cattle and act as windbreaks or shelterbelts. Social forestry also involves the participation of people on a large scale, so that natural forests are not destroyed. It helps create plant resource genetic banks, nurseries, recreation gardens, and so on. Although a few years ago monoculture (culture of one species of trees) was practiced for social forestry, polyculture (growing more than one tree species in the same land) is now considered better and practiced. Polyculture is based on indigenous or exotic species of plants which can provide not only timber but also much required fodder and fix nitrogen to increase soil fertility. The following species of plants are now widely used for social forestry: Acacia, Lukens, (Subabul), Prosopis (Jand), Sesames (Agastha), Casuarina, Tectona (Teak), Dalbergia (Shisham), Moringa (Shajan), and) Azadirachta indica (Neem) in rural India. Eucalyptus plantation was once considered a panacea for our economic ills. It is a quick growing tree which is used in the paper pulp industry. But it is not very afforestation suitable as firewood or timber, nor can it be used as fodder or for consumption of fruit. The eucalyptus has lost the popularity it once had.

Social forestry also aims to provide a good income to the farmer. Therefore plants which grow faster and reach a marketable size in 3-

8 years should be chosen. A modified version of social forestry is called agro forestry. Woody perennials are grown on the same land management units as annual agricultural crops and farm animals. These are grown either simultaneously or sequentially and the purpose is to maximize output on a sustained basis. Community forestry/ village forestry/gramya jungle is practiced around villages and panchayats. The three terms, social forestry, agro forestry and community forestry have much in common and are considered more or less the same thing. The social forestry programme has made the people conscious of the importance of afforestation and is gaining in popularity. Compensatory afforestation is stipulated by the Indian Government, while approving proposals for dereservation or diversion of forest

uses. Compensatory afforestation is insisted over equivalent area of non-forest land. The purpose is to compensate the loss of forest cover due to its diversion so that the net area under forest cover remains the same. If non-forest lands are not available or is available in less extent to the forest area being diverted, compensatory afforestation may be carried over degraded forests twice in extent to the forest area being diverted or to the difference between forest land being diverted and available non-forest land, as the case may be. The non-availability of non-forest land for compensatory afforestation is however, accepted by the Central Government only on the certificate from the Chief Secretary to the State/Union-Territory Government to that effect. However the compensatory afforestation has not been practiced vigorously and requires effective management strategy to support rural economy and livelihood options.

Agriculture

Agriculture is the main source of livelihood in villages. Agriculture employs 73% of the main work force in the country, although contribution of agriculture to NSDP has declined from 67 per cent (1951) to 30 per cent (2002).Small farmers and marginal farmers (owner of <1 acre of land) constitute about 84% of the total operational holdings with 50% of the agriculture land. Table 3 provides data on the agricultural production and productivity in the country (Prasad, 2009).

In the last 5-6 years in most food grain crops the production has reached a plateau and in some there is decline (Prasad, 2009). The rice productivity is less than 2 tons per hectare in most of the states. The productivity in Punjab, Andhra Pradesh, Haryana, West Bengal and Tamilnadiu is higher than the national average of 2.07 ton per ha.The population growth in the country is >1.3% and hence the production and productivity of food grains are to be enhanced. Prasad (2009) suggests many useful methods and the fact is that India needs another green revolution to cope with the demand of growing population.

Animal resources

India's fish production data is given in Table 4. The total production was about 7.31 million tons in 2007, out of the total fish production, aquaculture amounted to about 3.355 million tons and the capture fishery (both marine and fresh water) amounted to 3.95 million tons. The per capita fish protein consumption is not much and there is good scope to increase the aquaculture facility and capture fishery considering 7500 km of coastline and a huge continental shelf.

Productivity. The animals are important for providing organic manure for agriculture, dung-fuel for rural homes, and by-products like bone meal etc as manures etc.

However with reduction in grazing land and monetary constrains for stall feeding by the small and marginal land owners, government subsidy and Fish, prawn, and turtle form part of the livelihood support in Indian villages. Some of the ST and SC people also consume frogs, molluscs etc. The area under ponds, tanks, beels and fresh water lakes etc amounts to 3.42 million hectare. There is need of renovating village ponds for water harvesting and fish culture. The length of rivers and canals is 171334 km.The cleaning of rivers and canals for fish culture will boost the village economy.

The animal resources are utilized for different purposes and can be grouped under (i) milk group comprising cow, buffalo, goat and some minor contribution from camel and sheep, (ii) draught group comprising bullock, buffalo, horses, ponies, mules, camels etc, (iii) meat supplier group comprising buffaloes, cattle, sheep, goat, pig, rabbit etc, (iv) wool group comprising largely sheep, (v) egg group comprising fowls, and ducks. Marginal and small land owners are the owners of 71% of cattle, 83% of buffaloes, 88% of small ruminants, 70% of pigs and 74% of poultry (government of India-publication,

2005).In view of this the village livelihood sustainability is dependent on animal resources. As per the statistics, India produced about 89.35 million tonnes of milk (96% from buffaloes and cows, and 4% from goat), about 39092 million numbers of egg (20012-2002) and 4.91 million tonnes of meat including poultry meat (2001-2002) and the estimated annual growth is 5%.The live-stock products contributed 5.51% to the gross domestic product of the country in 1999-2000.The live-stock contributes about 25% of the agro-sector of the country in terms of money. This is the important sector for scientific research to increase the milk, egg and meat production. There is good scope to enhance this production and other welfare measures are required. Genetic selection of animal breeds, feed and nutrition aspects, health and disease aspects and opening of veterinary dispensaries and hospitals in villages are the priority areas. India is very rich in domestic animal resources and their management is the most important aspects for sustainable rural economy.

Rural Livelihood options

The rural livelihood options in India have centered around on agriculture, collecting fire wood, sal (Shorea robusta) or broad leaves (Tectona grandis) for making plates, kendu leaf for making cigarettes (bidi), Sal and Neem seeds for oil extraction, from nearby forests/ Gramya jungle, fishing in ponds and rivers, and in sea in coastal villages, vegetables from home garden, meat, egg, and milk from domestic animals and (cattle, buffalo, goat and sheep), poultry etc from bioresource-based cottage industries (bamboo, cane and wood for basket and furniture making) and small entrepreneurship (tea stalls, hotels, bakery and preparation of pitha, kiosks, cycle-motor cycle repair, tailoring, small laundry) etc.

The environmental liability of these livelihood options is either low or medium, but never assessed or looked into. My association in some World Bank funded projects, (Government of Orissa Project, 2007) field work for EIA report preparation for some industries in Angul and Denkanal districts, some coastal districts and in some districts of western part of Orissa is summarized below.

The non-farm activities center on goatery, poultry, dry fish trading, diary, rearing of wild boars, and piggery, and the forest based activities are leaf plate making, basket making, kendu leaf collection, firewood collection etc.

Environmental Sustainable Indicators

The forest provides firewood, house building material, fodder, fruits, medicinal plants and other utility materials to the villages with some control by village Panchayat.

The runoff water from the slopes of the mountain forest/Gramya jungle is stored in the water harvesting village ponds. These ponds provide surface water for the use of the villages throughout the year as some ponds do not dry up fully in summer. In view of this the village forest forms the life line for the villagers. A garland drain connecting to water harvesting ponds is usually constructed. The villagers depend upon the buffaloes and cows for milk, goat and sheep, poultry for meat and eggs and use the bullocks and buffaloes in the crop field for ploughing. The goat rearing is a profitable business for many village households.

The village forest and grasslands are also used extensively for grazing of domestic animals of the villages. In the changed situation of industrialization, the reduction/complete loss of grazing lands, villagers have problems of providing grazing land and unable to afford stall feeding. Domestic animals provide livelihood support to the people. However, employment opportunities for villagers in the changed scenario may increase their income, which will help to partially stall feed the animals.

Cropping pattern: Paddy, pulses and vegetables are the main crops. Pulses are grown followed by paddy. Brinjal, tomato, leafy vegetables, ladies finger, and vegetable banana are widely grown. Mango, Banana & Gauva are the major fruit plants grown by the villagers. The home gardens are also productive resources and are to be managed well.

The core poverty group is the landless wage earners, marginal and small farmers, (less than 1 acre land holding), ecosystem people (Dasmann, 1988; Gadgil and Heda, 2009) (forest dependent and other bioresource dependent- largely SC, ST, and some OBC). Poverty is chronic as income is erratic (Table 3).The livelihood pattern in villages vary according to cast, ethnicity and financial condition of the families. Since the livelihood pattern is largely bioresource based, government schemes should protect these resources and funding should be done to enhance bioresource productivity and management. The income from these occupations is subsistence level and do not meet the family expenditure. In view of this migration to urban centres in the state or to distant places in search of work to augment family income happens.

Environmental Aspects

Most of these livelihood options cited above have some environmental liability for people and these are never addressed in Indian situations. Table 4 gives an account of such environmental liability.

In coastal belt industrial pollution has affected growth of coconut, Bamboo, Kaju production largely due to SO_2 pollution.

Environmental Impact Assessment (EIA)

Since 1994 EIA study has become mandatory for all new projects or expansion or modernization of existing projects which have substantial stake on environment. As per 14 September 2006 notification of Ministry of Environment and Forests of Government of India, the generic structure of environmental impact assessment document and check list of environmental impacts have been stipulated. The purpose is to assess the expected impacts and to address them through an environmental management plan if the developmental project is allowed to operate. The check list includes land environment, water environment, vegetation, fauna, air environment, aesthetics, and socio-economic aspects, building materials, energy conservation and environmental management plan. The check list is a reflection of livelihood options prevailing in India. However there is a big gap between the EIA stipulations and implementation. The sustainability concerns of Indian villages have not been adequately addressed. The interest of multinational companies and big industrial houses of the country have received more attention than the natural resource management or furtherance of traditional livelihood options conserving the 'traditional ecological knowledge' available in village systems. Ramakrishnan (2009) points out emphatically that 'very traditional societies living close to nature and natural resources around them may have to have traditional ecological knowledge being brought in to a much larger degree so as to avoid social disruptions setting in, compared to the more modern societies who may need traditional ecological knowledge only to be brought in so as to create buffering mechanisms within the socio-ecological system and thus cope up with the ill-effects arising from excessive use of energy-intensive technologies.' I have had the opportunity of getting involved in some World Bank funded projects and projects of big industrial houses to understand socio-ecological systems in some villages of Orissa. My understanding is that sustainable development in Indian villages centres on creating facilities to enhance the existing livelihood options, adopting effective community participatory approaches, and sustainable management of bio-resources. This does not prevent opening of new avenues which will have minimum environmental impacts and can be addressed by application of science and technology and will create livelihood options of youth. The judicious implementation of EIA provisions is need of the time to address environmental uncertainties and to conserve bio-resource based traditional ecological knowledge and rural livelihood options.

References

Country Report on Animal Genetic Resources of India. 2005. Department of Animal Husbandry and Dairying, Ministry of Agriculture, Government of India, New Delhi.

Dash, M. C. (editor) 2007. The State of Environment Report-

Orissa, Government of Orissa, (Orissa State Pollution Control Board,

MOEF, Government of India, Academic Staff College of India, Hyderabad funded), Bhubaneswar. 283 pages.

Dash, M. C. 2007.Terrestrial Ecology Study in Jharbandh-Galapada-Tarkabeda village Areas of Denkanal District for the Proposed Iron-Steel Industry By Rungta Mines Ltd., Submitted to Government of Orissa as per Orissa Resettlement and Rehabilitation Policy, 2006 (Funded by Agricultural and Rural Development Consultancy Society, Bhubaneswar, 61 pages.

Dash, M. C. 2009 .Environmental Status and Sustainability Report As per Orissa Resettlement and Rehabilitation Policy, 2006, Prepared for ARDCOS, Bhubaneswar (LancoBabandh Power Private Limited, Denkanal, Orissa), 73 pages.

Dash, M. C. and Dash, S. P. 2009. Fundamentals of Ecology, Tata-McGraw-Hill Publication, 562 pages.

Dash, M. C. and Kar, C. S. 1990. The Turtle Paradise: Gahirmatha (An ecological analysis and conservation strategy), Interscience Publishers, New Delhi. p. 293.

Dasmann, R. 1988. Towards a Biosphere Consciousness. Pages 177-188, In: Worster, D. (Editor).The Ends of the Earth, Cambridge University Press, Cambridge, U. K.

Gadgil, M. and Heda, N. 2009.First Prof.R.Misra Birth Centenary Lecture-Of Rivers, Fish and Poison. Internat. J. Ecol. and Env. Sci. 35 (1) : 1-11.

F. A. O. 1985. Year book of forest products, Rome.

Government of Orissa-Integrated Social and Environmental Assessment Study, Prepared by Varun Techno Infrastructure Pvt. Ltd, World Bank Project) 2007. Volume 1-3.

Pattanaik, A. 2006. Coastal and Marine Ecosystems, pages 228-259, In 'State of Environment Report' (M. C. Dash, Ed.), Orissa State Pollution Control Board, Bhubaneswar.

Prasad, R. 2009. Food Production in India-Progress and Future Stategies, Proceedings of National Academy of Sciences, India. Section B. 79: 104-109.

Ramakrishnan, P. S. 2009. Towards Addressing Societal Concerns: Moving Through Genecology and Ecosystems to Socio-Ecological Systems. Internat. J. Ecol. and Env. Sci. 35 (1) : 13-33.

Sharma, A. 1987 Resources and Human Well-being: Inputs from Science and Technology, Presidential Address, Indian Science Congress, 74th Session, Bangalore, Indian Science Congress Association, Calcutta.

State Forest Report. 2005. Government of India, New Delhi.

SPAN, (All issues) 2006. American Center, 24 Kasturba Gandhi Marg, New Delhi-110001

World Wide Fund for Nature.2009

8

Conservation and Management of Natural Resources: An Overview

Introduction

A *resource* is a means for satisfying human and social requirements in a given space and time. Resources are basically of two kinds, (a) renewable, and (b) non-renewable. Renewable resources, such as plants (crops, forests, etc.) and animals (milk and meat producing) are replaced from time to time because they have a life-cycle, and a continuous harvest is possiblc. Some resources do not have a life-cycle but can be recycled. These resources, like water, are also classified as renewable. Natural resources are also classified as (a) inexhaustible, and (b) exhaustible. Resources which are usually not changed by man's activities, and are abundantly available and expected to be available for millions of years, are said to be inexhaustible. Solar energy, atomic energy, wind power, power from tides and so on is classified as inexhaustible. Renewable resources can also be classified as inexhaustible because they can be managed properly and renewed. Renewable resources such as plants and animals become extinct if not managed properly. They are maintainable

Resources which are not regenerated because they do not have a life-cycle or are not recycled are termed non-renewable. Mineral deposits are formed slowly over millions of years and once used cannot be regenerated, e.g. fossil fuels, such as petrol and coal. These resources are classified as exhaustible because their deposits on this planet are finite. Purely exhaustible resources like fossil fuels and some minerals are non-maintainable.

Nonrenewable Resources

Mineral Resources: (Coal, iron ore, bauxite, chromites, etc and their uses) : A number of minerals, such as coal, iron ore, chromites-chromium, copper, bauxite-aluminum, petrol, etc. are essential for our industries. India has a thriving mining industry with mineral sector contributing about 5-7% of gross domestic product. The number of minerals mined in India can be grouped into (i) fuel minerals like coal, petroleum, natural gas, lignite, (ii) metallic minerals like iron ore, chromium ore, bauxite etc. and (iii) non-metallic minerals like lime stone, dolomite, phosphorite, clay etc. India produces 89 minerals including four fuel minerals, 52 non-metallic minerals, 11 metallic minerals and 22 minor minerals. India stood as world's largest producer of mica blocks & mica splitting and ranked second in the production of chromites, third in coal & lignite, and barites, fifth in iron ore, sixth in bauxite and manganese ore, eleventh in aluminium and twelfth in crude steel in the World (IBM, 1999). With its contribution of about 16 percent in exports and 20 percent in imports, the mineral sector is an important component of India's foreign trade.

Fuel minerals contribute 88% of the total value of mineral production (coal 42%, petroleum 33%, natural gas 10.5% and lignite 2.5%). The metallic minerals and the non-metallic minerals account for 6% each of the value of the minerals produced. India possesses about one-fourth of the world's iron-ore deposits located largely in Jharkhand and Odisha. It now produces more than 100 million tonnes of iron ore annually. It has been estimated that India has about 1, 250 million tonnes of bauxite deposits, located in Jharkhand, Chhatisgarh, Odisha, Andhra Pradesh, and Gujarat. Alumina is extracted from bauxite and alumina is used to produce aluminium. About 80 million tonnes of manganese deposits are also found in India with Odisha as the leading producer state. Out of 193 million tonnes of chromites deposits, about 140 million tonnes are found in Odisha. Copper deposits are scanty and found in Jharkhand, Bihar and Rajasthan. Gold deposits are meager; India also has good reserves of uranium and thorium in Jharkhand, Bihar and Rajasthan. These are important for the generation of atomic energy. India is rich in some 35 minerals. About 6% of the value of minerals production is due to metallic minerals.

It has been estimated that the coal reserves of India amount to about 86, 000 million tonnes. At present more than 575 coal mines are present in India. Annual production is more than 400 million tonnes and consumption is more than 500 million tonnes. The annual production of petroleum is more than 10 million tonnes and our reserves are located in the valleys of the Brahmaputra, the deltas of Cauvery, Krishna, Godavari and Mahanadi rivers, coastal areas of Kerala, Andhra, Odisha and offshore sites near Mumbai and so on.

Renewable Resources (Forest, Grasslands, Aquatic Resources etc & Management Strategy)

Renewable resources are of conventional types, such as forests, crops, other plants, wildlife, and aquatic resources, and non-conventional, such as solar energy, wind energy, wave and tidal energy, biogas, atomic energy, and so on. These resources can also be grouped as water resources, energy resources, agriculture, rangelands, forest resources, wild life, aquaculture, etc.

Water Resources:

We have already discussed the importance of water in the cycling of materials. Water is needed for daily use by organisms, for irrigation, navigation, industrial use, electricity production and domestic use. Of the total water resource of the earth, 97.3% is salt water and the rest fresh water. This 2.7% amounts to about 1.4 billion km^3 of water.

Of this, about 77.2% is permanently frozen, 22.4% occurs as ground water and soil moisture, 0.35% is found in lakes and wetlands, and 0.01% in rivers and streams (World Resources, Basic Books, New York 1986). The total volume of ground water found in underground reservoirs, called aquifers, is estimated to be $42.3 \times 10^{10} m^3$. At present about 25% of the ground water is being used by man. Agriculture uses the maximum amount of water in the world. These amounts to about 73% and leads to a lot of pressure on ground water. Excessive use of ground water depletes aquifers, lowers the water table and may lead to Stalinization, water logging and alkalinisation of soils. As the population goes on increasing, so do agriculture and industrial activities. Thus there is considerable demand for water resources. It has been estimated that the fresh water needs (drinking and food) of a person are about 3 litres per day. This does not include the water requirement for other purposes, as per the present population of the world; > 6 billion m^3 of water is required per year. But the distribution of fresh water is not uniform in all countries and the pattern of use also varies. Therefore, there is a scarcity of fresh water in many parts of the world. The use of ground water at rates higher than those at which it is recharged should be avoided. Water pollution should be avoided and steps taken to store runoff water. One estimate shows that about 1.3 billion people in world do not have access to clean drinking water.India receives about 3 trillion m^3 of water from rainfall, which amounts to about 105 to 117 cm annually. This is a huge water resource and the largest in the world. But almost 90% of this precipitation falls between mid-June and October. Fourteen major river systems, such as the Ganges, Narmada, Brahmaputra, Mahanadi, Cauvery, Krishna and Godavari, account for 85% of surface flow and share 83% of the drainage basin. They serve 80% of the total population. There are about 100 medium and minor rivers. The storage capacity is 3.65 million m^3. Of the total annual precipitation, India utilizes only 10%, which may increase to about 26% by 2025. The ground water is also not well tapped except few states. We need to apply improved technology to increase water harvesting and storage capacity and ground water recharge and utilization.

The Central Ground Water Board has estimated that the available ground water in India is 210 billion m^3 and the annual utilization potential is about 42.3 million hectare, only one-fourth of which is used at present. However, the ground water availability is not adequate in provinces like Tamilnadu and Andhra Pradesh. Due to large scale deforestation and monsoon failure, there is a regular occurrence of drought in Kalahandi and some other districts of Odisha, Karnataka, Rajasthan, and Maharashtra. The supply of drinking water in Indian villages is not adequate. There is no organized water supply in many villages and even in some small towns. Although there is a lot of precipitation in India, sufficient care is to be taken to manage this water efficiently.

Energy Resources

Non-renewable energy resources like oil, gas and coal account for about 90% of the world's commercial energy requirement (petrol 45%, natural gas 20% and coal 25%. They are now in short supply and also generate a huge amount of environmental pollutants when burnt.

The conventional sources of energy are mainly: (i) Biomass or dried organic matter (renewable) : These are, however, renewable sources of energy. Fuel wood consumption provides about 43% of the total energy/ consumed in all developing countries taken together but this amounts to about 14% of the total energy production of the world. Fuel wood demand in India was about 150 million tons in 1985 and this has doubled in 2005 AD. As natural forests are depleted, they will be able to provide 50-75 million metric tons and the rest of the demand will have to be met with fuel wood plantations, which require lot of effort and sound management strategy. (ii) Fossil fuels (nonrenewable) Fossil fuels are non-renewable source of energy although these are still forming very slowly as compared to their consumption.

Concept of Demographic Quotient:

Could (1969) gave the concept of demographic quotient (*Q*), which is defined as

$$Q = \frac{\text{Total resources available}}{\text{Population density}} \times \text{per capita consumption}$$

If the value of the quotient goes down, so does the quality of modern life. The demand for energy doubles every 10- 14 years and in modern times energy consumption is taken as an indicator of a country's development. India, with about 16% of the world's population, consumes only about 2% of the world's energy production while the USA with 6.25% of the world's population, consumes 33% of the world's energy production. In India, 80% of the population depends upon dung, wood and agricultural waste as the main energy source.

Due to the indiscriminate felling of trees, forests have been depleted and fuel wood has become scarce. Cattle dung, which could have been used for agriculture, is used as fuel.

Solar energy: It is the sole source of energy for the biological world. Since India receives abundant solar energy, it is advantageous to develop photovoltaic devices to harness this energy. In India with an average of 250 clear sunny days in a year, such systems have proven viable and cost-effective. The efficiency of conversion of sunlight into electrical energy is up to 15-22 percent at the present level of knowledge of photosensitive materials and technology. Advances in molecular biology and nanoscience may provide promise for a break through. At present (2005) Japan, Germany & Spain generate respectively 2422 MW, 1500 MW and about 50 MW energy from solar source. India's target is 20, 000 MW installed capacity by 2020.

Wind energy: Windmills convert wind energy into mechanical energy for raising water from wells and rivers to the ground surface. Alternatively, wind energy can be converted into electricity to supply a small town or village. Windmills are in use in many countries, notably Holland. At present wind energy accounts for more than 50 per cent of the 8000 MW energy produced through renewable sources in India. This has great potential in India, as there is a long seacoast. In 2005 wind energy generation in Germany, Spain & USA were respectively 18428 MW, 10027 MW & 9149 MW.

Wave and tidal energy: Waves and tides can be converted to electrical energy. In many hilly regions of India flourmills are run by water energy. Hydroelectric power is generated if a natural or artificial waterfall is created to turn a turbine (comparable to a pedal wheel) rotating fast. Areas where rivers flow into the sea experience waves and tides, and electricity can be generated thereby. India, having a large coastline and major river systems and mountain ranges through which these rivers flow has immense potential for generating electricity from waves and tides.

Geothermal energy: In hot springs we find superheated steam. In India there are 46 hydrothermal areas where the water temperature exceeds 150°C. These hot springs can be utilized to generate electricity. The superheated steam can be used to run machines, turbines, and so on.

Biogas: Cattle dung is used to produce biogas. India has 185 million cattle, 61 million buffaloes, 45 million sheep, 97 million goats, 1 million horses, 1 million camels, and 1 million other livestock. These animals produce a huge quantity of dung, which can be utilized for biogas production. Water hyacinth, *hydrilla*, duckweeds and algae can act as supplements for production of biogas. Besides cooking, biogas can be utilized to produce steam, which can be used to run machines and turbines for generating electricity. Left-over slurry can be utilized as manure. India has appreciable potential for biogas energy (see Environment and Resources-4, Foundation Course in Science and Technology, Indira Gandhi National Open University, New Delhi, 1989).

Some 1, 500 to 2, 000 million tonnes of wet animal dung is available per year in India and at a 66% collection rate, 25000million/ m^3 of biogas can be produced. The slurry can produce > 200 million tones of organic manure per year, equivalent to 1.5 million tonnes of nitrogen, 1.5 million tonnes of phosphate and 1, 0 million tonnes of potash as fertilizers. At present the amount of dung production is much more. There are about 0.7 million biogas plants in India now.

Biomass Energy: Some conventional energy resource, such as biomass fuels (wood, etc.), are also renewable.

The present (2000s) firewood demand in India is more than 350 million tonnes. Biomass such as firewood, agricultural residue, bagasse, crop stalks, coconut shells, waste from agro based industries, and animal dung can be used to produce power. Direct burning of these wastes is inefficient and leads to pollution. If these wastes are combusted in a gasifier at low oxygen and high temperature, can be converted into a gaseous fuel called producer gas. This gas, although has a lower calorific value compared to natural gas or liquefied petroleum gas, can be burned with high efficiency and without emitting smoke. India generates >600 million tonnes of agricultural residue per year. These wastes can generate 79000 MW of power if all of this waste is gasified (SPAN, 2006). This energy will form more than 60 per cent of the total power available in India from all sources. It is feasible to set up a biomass based power plants of 10 to 20 MW in every village block, consisting 100 villages.

Atomic energy: Atomic reactors are utilized to harness energy from the atom in a controlled manner. Nuclear reactors produce heat, which is used to produce steam. One kilogram of natural uranium (written as U^{238}) generates energy equal to that generated by 35, 000 kg of coal. In India the nuclear energy capacity was only less than 4 % of the total installed power generating capacity. The plan is to produce about 10% of the total energy needs of the country from nuclear source by 2020. An hour of coal-generated 100-watt electric light creates a half-pound of atmospheric carbon, a bucket of ice makes a third of a pound, an hour's car ride 5 pounds.. Atomic power is less expensive than it used to be – but could it possibly be cost-effective? What is a rapidly carbonizing world to do? The high-minded answer, of course, is renewable and nuclear energy. *(Source*: SPAN: 2005 & 2006 & other sources).

Forests as Renewable Resource

Forests constitute 90% of the global biomass. Forest ecosystems are great resources, since they provide habitat for wild life, fuel wood, fodder, fibre, fruit, herbal medicines, timber, and several raw materials, which are primarily used in wood-based industries, including paper and pulp industries. Forests regulate climatic conditions, such as rainfall, humidity and temperature regime of the area and protect the soil and landscape from wind and erosion. Thus they have three broad functions, (a) *productive*, (b) *regulative*, and (c) *protective*.

A climax forest may account for 400 ton of dry matter per hectare, equivalent to about 22 to 156 x 10^{10} cal per hectare. The large animal biomass may account for about 1, 000 kg per hectare. A large amount of mineral substances are stored in the vegetation, particularly in tropical forests. It has been estimated that some 2 ton to 53 ton of minerals per hectare are accumulated in tropical forests. The long-term stability of the forest ecosystem depends upon the input and output of nutrients.

Livestock as Renewable Resources

Livestock are important resources, as they provide milk, eggs, meat, skin (leather) and other products. The total livestock in India mount to a huge population consisting of more than 190 million cattle, 118 million goats, 61 million buffaloes, 45 million sheep, 1 million horses and ponies, 1 million camels and about 1 million other livestock. Besides, there are 156 million tonnes poultry, fowls, ducks, turkeys, etc. They provide about more than 15 million tonnes of meat, 112 million tonnes of milk, 40 million kg of wool and 13 million eggs.

The amount of dung produced is huge and is used for biogas production and as manure. Livestock are also used for rural transport and agriculture. To get a sustained yield, their proper breeding, health care and management and diet are essential. Recent advances in biology have helped produce better livestock. The total milk production has increased from 17 million tonnes to 100 million tonnes in 2006-07.

Livestock contributes at least 40, 000 million rupees a year to our national economy. Of this, poultry contributes about Rs.10, 000 million. Almost all the world's fowls trace their origin to India's red jungle fowl (Gallus gallus). Ducks constitute about 10% of the total poultry in India.

Aquaculture as a Renewable Resource:

Fish are an excellent source of protein. They feed on plankton (phytoplankton and zooplankton), weeds and organic matter. Crabs, prawns, mollusks, etc. are also aquatic organisms and are a very good source of animal protein for human beings. Over half the population in the developing countries gets at least 40% of its animal protein food from fish (FAO, Status of Food and Agriculture, 1984, Rome). The natural pearl industry depends upon oyster culture. Fishery includes capture fishery and culture fishery. Capture fishery is usually practiced in oceans, large lakes and other aquatic ecosystems. Culture fishery is usually practiced in inland water bodies and artificial ponds.

The global marine fish catch was about 76 million tonnes in 1984 and is on the increase. India's marine fish catch was about 1.2 million tonnes in 1984 and it has increased to 3.1 million tonnes in 2005. India has a very long coastline (about 5, 600 km) and a continental shelf of about 2, 59, 000 km^2. Thus there is a vast area available for tapping. Marine resources include the Bay of Bengal, Arabian Sea, and Indian Ocean, numerous gulfs, coral reefs and mangrove swamps. The estuaries of major rivers, backwaters, lagoons, and brackish water lakes like Chilka, add to India's marine resources. They form an additional 2.6 million hectares. There is a good possibility of increasing the fish catch along the Indian coast using the sophisticated scientific techniques now available. In 1984, India produced about 3 million tonnes of fishery products, occupying the eighth position in the world. It has increased considerably and it has been estimated that this production can be increased to about 12 million tonnes per year. However, fishing remains confined to a narrow coastal region of 16 km and the rich offshore and deep waters remain poorly exploited. India has about 1.6 million hectares of inland waters. It has many major river systems, such as the Ganges, Jamuna, Brahmaputra, Narmada,

Mahanadi, Cauvery and Krishna. Besides, there are canals, ponds, lakes and irrigation channels where culture fishery can be practiced.

The average aquaculture production is about 600 kg per hectare per year in India. This yield can be increased to 11 tonnes per hectare per year using composite fish culture (using surface, column, and bottom feeding fish in the same pond and thus avoiding competition among them for food). The yield of air breathing fish in swamps is now 15 tonnes per hectare per year. The total quantity of aquaculture production in the country in 2015-2016 is about 10 million tonnes

Conclusion

The renewable resources, especially the Bioresources are threatened. The recent thrust on industrialization in culturally rich and resource rich but economically poor states has created a situation where traditional livelihood options, which are Bioresources based, are threatened.

In view of this Environmental Impact Assessment (EIA) and Environmental Management Programmes- implementations is very important.

India is a store house of traditional knowledge and this is to be vigorously promoted and protected from plagiarism. Documentation of traditional knowledge is very important and an area of innovative research.

The integration of science and technology, traditional knowledge, and judicious application of EIA protocol can make significant contribution to protect resources for linking to livelihood options for sustainable development.

References

Dash, M.C.2012. Basic Concepts of Environmental Management for Sustainable Development, I. K. International Publishing House, New Delhi, 320 pages (In Press).

Dash, M.C. and Dash, S.P.2009.Fundamentals of Ecology, Tata-McGraw-Hill Publishers, New Delhi, 562 pages

9

Wildlife Conservation: An Over View of Existing Laws for Their Protection

Introduction

India is one of the mega-diversity nations in the world. The Western Ghats, Northeast India-Eastern Himalayas and the Andaman- Nicobar islands are globally recognized wilderness areas of high endemism in flora and fauna. They are designated as hot-spots owing to the richness of endemic biological diversity and also because of the high degree of biotic threat to the same. Approximately 18, 644 species of higher plants and 89, 451 species of animals have been reported from India (Chavan et al. 2004). The faunal and microbial diversity exceeds 50, 000 species. The number of species is not uniformly distributed in the region. The species numbers are more in the Western Ghats, North east –Himalayas and Andaman Nicobar islands. However, other areas such as Western & Central Himalayas, Easternghat, sea coast and deserts have their own unique species of plants and animals. India has enacted many laws to protect the environment and the wildlife (Mohanty, 2011).India is also a signatory to the Convention of Biological Diversity (CBD). However, there are problems in implementing the laws in true spirit, and approach to address the conflict between development and environmental protection is not always scientific and rational and other considerations influence decisions. India has enacted laws to protect biodiversity and wildlife and is signatory to CBD. It appears that there was lack of coordination between different government departments before enactment of these laws and also during implementation due to some inherent constraints. Provisions of the forest acts, wildlife act, biodiversity act and EP act and the constraints have been discussed.

Examples of endemics in India and threat to them

The Asiatic Lion: once seen across the Indian grassland plains are now limited only to a single forest area, - The Gir forest of Gujarat. The habitat is threatened by mining around the Gir forest area. It is alleged that the conflict between mining mafia and the environmental activists opposing it has led to the shooting of Amit Jethwa (environmentalist) by unknown gunmen in 201

INDIAN ENDEMISM

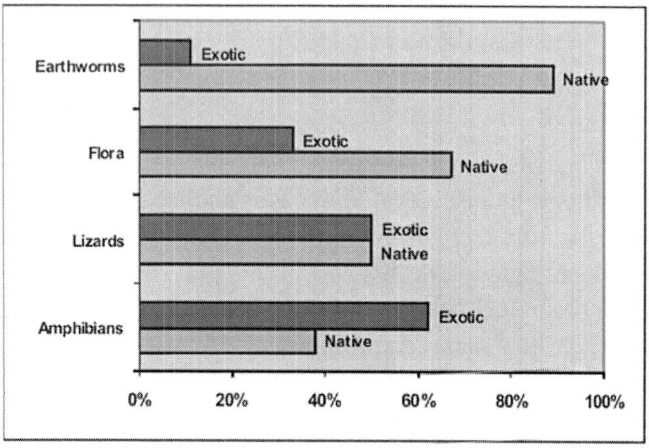

Nepenthes khasiana: the carnivorous pitcher plant is only present in Meghalaya state. Indiscriminate coal and limestone mining in the region is threatening its habitat. *Sengai*: The Brow-Antlered Deer is known only from a highly specialized wetland ecosystem in Manipur. It is under threat from pollution of the water bodies and loss of habitat. *Aponogeton satarensis*: a plant growing in monsoon puddles is seen only at 5 locations in Maharashtra; in Mountains at altitude 1100m.Ecosystem fragmentation is the cause. It is seen only in an area of about few square kilometer. Figure 1 (Julka, 2010) gives an account of endemism in India.

International support has also focused on protecting Indian wildlife and wilderness. Some of the major initiatives have been the Tiger, Elephant, Crocodile & Sea Turtle conservation projects ongoing across India in many states. Man-Animal conflict, Weak governance, Lack of adequate awareness, Cultural conflict etc are the cause. Several policy decisions have been taken to conserve the Indian biodiversity. Ministry of Environment and Forest (MoEF) is functioning as nodal agency with decision-making power about the Indian biodiversity and environment.

Protected Areas

The recent approaches to conservation have focused on conservation of species as well as/ ecosystems/ landscapes in which species inhabit. e.g. WWF's species and landscape programme for tiger and Indian elephant. Conservation policy has been designed to develop "Protected Areas" (PA), which is established by legal procedures. The threatened vertebrate species in India are ~ 22 mammalian, 15 Aves, & 17 Reptiles. The Environment Impact Assessment (EIA) document -1994 and its amendment from time to time, latest being 2011 (under EPACT-1986) should have taken care for establishment of new industries in proximity of PAs without damaging integrity of PAs. Is it happening? The Biodiversity enumeration & conservation aspects have lacunae. Under EIA and EMP the flora (vegetation) and fauna enumeration and quantification have not been adequately defined.

The state forest departments usually give importance to trees only. Shrub, herb etc do not receive much attention. They do not give importance to ecological diversity & only mention taxonomical diversity & do not quantify the biodiversity. The wildlife departments of the government usually study the large-sized fauna such as mammals, and Limited number of species of Herpetofauna- such as crocodiles & sea turtles. They do not enumerate SOIL BIODIVERSITY which is ecologically very important. University-biology faculty and individual biodiversity experts should be involved. The enumeration profile and level of quantification should be specified.

PAs are ecosensitive areas. As per EP-ACT, no industrial developmental projects, which will have negative impacts on the environment, wildlife and their movement should be allowed to be established within 10km from the boundary of PAs/ecosensitive and critically polluted areas.. The project requires prior approval of the ministry of Environment and Forests. This is not a wildlife act or forest act. This is a Pollution Control and environment act. In view of this effective coordination between the Pollution Control Board and forest and Wildlife Departments are required. In practice in many places industrial developments have happened or being happening within 10 km area of ecosensitive areas.

Stakeholders are groups and individuals who affect or are affected by the activities of development projects such as mining, which happen in wildlife rich forest areas. The Stakeholders are: Ecosystem people, local communities; & people at large; government and other institutions (Regulating institutions) with an interest in or responsibility for the management or protection of natural resources, especially forest resources including wildlife; Mining companies, insurance companies, who may also impose environmental requirements or standards; conservation interest groups,, both international, national & NGOs, academic Institutions for Research.

The Wildlife Act

The term 'wildlife' as pcr the Wildlife Act means *'any animal, bees, butterflies, crustacean, fish and moths: aquatic or land vegetation which forms part of any habitat' {sec.2 (37) }*. The original Wildlife Act was enacted in the year 1972 and thereupon amended several times to further improve the objective of protection and the conservation of the wildlife.

The Act consists of 66 sections and 6 schedules. As the title of the Act suggests it mainly aimed at the protection of the wildlife but with the amendment to the Act in 1991, specific species of plants were brought under the Act for protection. The main objectives of the Wildlife protection Act are: The State Government is given the power to declare the Sanctuaries National Parks aimed at the protection of the wildlife. Conservation and protection of the flora and fauna of the protected areas, Regulation of the hunting of the wild animals and the birds, Regulation of possession, acquisition or transfer of, or trade in wild animals, animal articles and trophies and taxidermy etc. Constitution of authorities to achieve the objectives of the act.

The Wildlife (Amendment) Protection Act, 2002

The changes that were brought about in the Act did make a considerable changes aiming towards the conservation of the wildlife in a better manner. The 1972 Act had the concept of 'Closed Areas', which was completely deleted in the 2002 amendment of the Act. Some features include: The hunting of the wild animals is prohibited by the Act except in special situations .Protection of Plants: The Act not only deals with the aspect of wild animal protection, it also concentrates on the protection of the plants. Penalty and forfeiture of property derived from illegal trade or hunting has been introduced.

Forest Conservation in India

Table 1 gives a history of forest conservation in India. During

Table 1. Historical aspects of forest conservation (source: many)

Year	Legislation/Control Remarks
1807	East India Company acquired royalty rights over Teak. Locals were not allowed timber for domestic use. Authority for teak felling and selling was done by executive order.
1846	Sanction Authority extended to all forests and forest produce. Locals including forest dwellers were not allowed for extraction of timbers from all forests
1857	Sepoy mutiny (forest dwelling people provided safe hiding places to rebels). Company administration prohibited and withdrew all access rights and privileges to fuel, fodder and other minor produces.
1860	Company's sovereignty extended to total area of forest land.
1864-	Imperial Forest Department was formed Government Control over forests and forestry was strengthened. The minor produce etc became inaccessible to forest dwellers.
1865, 1878, 1927	Series of forest Acts Promulgated. In order to legitimize Government Control,
1927	Act continued after independence as forest legislation. Rights of Forest dwellers (Tribal people) were not recognized.
1947-1970	Commercial exploitation of forest land for mining, other industrial activities & for creating farmland for large peasantry class were allowed. Forests were degraded, wild life poaching was rampant and became luxury activity of upper echelon of people.
1970-1988, 1972-	Wild life protection Act was enacted.,
1980-	Forest Conservation Act was enacted, The role of forest dwellers, tribals were not given importance & their Rights were not recognized.
1988	National Forest Policy was formulated. Forests were recognized as Resource/local resource. Participation of local forest protecting communities became mandatory in regeneration of degraded forests from 1994.
1996-	Panchayat Extension to Scheduled Areas Act, (PESA) 2006- The Scheduled Tribes & other traditional forest dwellers Recognition of forest Rights) Act enacted

colonial period forest was used only for exploitation of resources and no attention was given to tribals and some other people who lived in the forest. Wildlife conservation did not receive much attention until 1927.

The Forest Conservation Act-1980

The Forest Conservation Act -1980 is considered to be one of the most important legislations as far as the conservation of the forests is concerned. The main objective of the act was the conservation of the forests and matters connected therewith, which was completely ignored by the Indian Forest Act of 1927. The Forest Conservation Act was amended again in 1988-89.Some of the important objectives of the Act are: Imposition of restrictions on dereservation of forests, Restriction on the use of forest land for non forest purpose, Conservation of the forest wealth Check further reforestation, Maintenance of ecological balance .

National Forest Policy 1988

The Concept of community participation in the protection and the maintenance of the forests was the highlight of this policy. Some key features are the Preservation and restoration of the ecological balance by environmental stability., Conservation of the natural heritage.;Fulfilling the basic requirements of the people dependent on forests like fuel, wood, fodder etc.;Protecting the customary rights of the people who are dependent on the forests; Recognition of the interrelationship between the tribals and the forests; It encouraged public participation and involvement of women in the management of forests.& encourages involvement of local people in management. It also encouraged a time bound Afforestation programme. It extended certain rights including grazing rights and deals with shifting cultivation, forest based industries, forest extension, education and research etc. An important development was the formalisation of the Joint Forest Management

Scheduled Tribe Act-2006-Forest Conservation

On one hand the Act provides livelihood security to Tribal people/ Forest dwellers (4 ha for family-75 years residency in forest land).On the other hand it has created problems for wildlife and forest conservation (wildlife corridor affected, core area disturbed), podu (shifting cultivation) practice is against forest conservation, Akhand Shikar kills wild animals).Core area should not have been disturbed. They should have been allowed to settle in peripheral area/buffer areas. This could have been solved through negotiation & creating awareness and providing good financial support. The enactment of the act was initiated by the Tribal welfare department but the implementation is done by the forest department and Revenue department. Wildlife conservation and wildlife movement/corridors are being affected. Implementation has immense problem. Again coordination between different government departments before enactment of the act and for implantation is must. But this has not happened and the implementation has not become very effective.

The Biodiversity Act 2002

The government of India in the year 2002 enacted this legislation. India is a signatory to the Convention of Biological Diversity-1992 (CBD) of Earth Summit. The enactment of Biodiversity Act of India has been done to meet the international obligation. It is designed as a complementary act that would go hand in hand with the other legislations like the Wildlife Protection Act, Forest Conservation Act and Indian Forest Act, etc.

Objective of the Act:

To Protect India's rich biodiversity and associated knowledge from being misused, especially by foreign individuals and organisations without sharing the benefits arising out of such use.

To conserve for sustainable use and equitable sharing of benefits.

To promote at all levels and to do documentation of biological diversity, its uses and associated knowledge. To develop a well-designed Biodiversity Information System for India to serve as the knowledge base.

To establish a three-tire management system i.e., a National Biodiversity Authority, State level Biodiversity Boards and local level Biodiversity Management Committees.

To Check Biopiracy

To regulate access to genetic resources and associated knowledge by foreign individuals and institutions and ensure equitable sharing of benefits arising out of resources and knowledge available in the country.

In short, the Act is aimed to protect and regulate access to plant and animal genetic resources and traditional knowledge (TK). A three-tiered system of regulation is envisaged under the Biological Diversity Act, which consists of the National Biodiversity Authority (NBA) at the head, followed by State Biodiversity Boards (SBB) and local-level Biodiversity Management Committees (BMC).

The act also specifies offences and penalties. The court can take cognizance over the complaints under this Act only if the complaints go through the central government. The offences that are committed under this act are cognizable i.e., the offender can be arrested without warrant and non-bailable warrants. The penalties fixed by the act have been fixed much higher than other acts.

The fine is one lakh rupees and in case of second offence it is two lakh rupees and for a continuing offence it is prescribed as two lakh rupees for each day of such disobedience or non-compliance. In case of commission of an offence by a company all those who are responsible for the management and conduct of the company will be held liable.

After two years of the enactment of the Biodiversity act, government notified the Biodiversity Rules. The rules were formally notified in the official gazette on 15th April 2004. The Biodiversity Rules are the executive orders made by the Government in order to carry out the purposes of the Act (Section 62). This is the time to review the loopholes if any and functioning of the act and the benefits that the country has received because of the enactment of this act.

Lacuna in Policy Implementation

In recent time the development issues and environmental protection issues have created conflicts. Since natural resource maps are usually not considered while taking decisions on wildlife conservation, this has created serious conflicts for their conservation and for creating wildlife corridors and for mining activities. Mineral resources are essential for industrialization and development. Biodiversity, especially wildlife are also very important for many reasons. In view of this very judicious decision considering cost-benefit analysis, analysis of ecosystem services and long term benefits should be taken. Political and other considerations should not be considered at all. Another issue is ecotourism, which should not be allowed in core areas of PAs. But this is not happening.

Human-Animal Conflict:

For Human-Sea Turtle conflict and sea turtle conservation, Dash, M.C. & Kar, C.S. (1990) suggested a management plan, for Human-Elephant conflict, Swain (2004) suggest many remedial measures, for Human-Crocodile conflicts, Kar & Patnaik, (1999) have given many suggestions, These are very visible; not been implemented fully and not solved the problems adequately, although compensation to people for loss of life & property is given by the Government. Habitat loss & fragmentation, & availability of food are the constraints.Controversy with regard to (i) wildlife census, (ii) carrying capacity studies and (iii) follow up actions continue.

Universities are known for wildlife study & research, especially mammalian & Herpetofauna- They should develop regional & species specific protocol for their conservation. These aspects require deliberations and convergence of views for effective action programmes.

References

Chavan, V., Watve, A.V., Londhe, M.S., Rane, M.S.Pandit, A.T. and Krishnanb, S., 2004.Cataloguing Indian Biota; The Electronic catalogue of known Indian Fauna. Current Science, 87 (6) ; 749-763.

Dash, M.C. and Kar, C.S.1990.The Turtle Paradise Gahirmatha: An Ecological Analysis and Conservation Strategy, Interprint Publishers, New Delhi, 295 pages

Julka, J.M. 2010 (personal communication)

Kar, S .K. and Patnaik, S.K.1999. Status, Conservation and Future of Salt water Crocodile in Orissa. Envis (Wildlife and Protected areas), Wildlife Institute of India, Dehradun, 2 (1) :24-28.

Mohanty, S.K. 2011. Environment and Pollution Laws., Universal Law Publishing Company, New Delhi, 853 pages.

Swain, Debabrata, 2004. Asian Elephants: Past, Present and Future, International Book Distributors, Dehradun, 226 pages.

Part III : Food Security, Environment and Energy

10

Food Security is the Call of the Time

Introduction

This year's message in the World Environment Day is 'Eat, Save-Reduce your Foodprint'.This slogan is linked to excessive consumerism, especially practiced by many industrially developed countries and by the rich people in all countries.Besides, there is wastage of food due to many reasons. The calories need per capita on the average may not be more than 2000 per day. These calories should come from a balanced diet. But people abuse the food intake and do not care much for others in the world. World's population is more than 7000 million and India's population is more than 1210 million. To provide food and fresh water and other resources to this huge population is a daunting task. The population is growing on the average @1.3% in the world and @1.41% in India. This paper discusses the different aspects of increasing food production, especially grain production, animal protein production, water availability etc for achieving food security in India.

Since the Rio-1992 summit, and adoption of Agenda-21 by the global community, most of the countries including India have set sustainability as a key goal for their development. It is generally accepted that the four pillars of sustainable development are the environmental, social, economic, and technological wellbeing, especially green technology (Figure 1). To achieve Sustainable Development (SD), different management systems are adopted.

The defining aspects of SD are: Population control, energy use, increasing productivity, food security, water resource conservation, and biodiversity conservation, development of Science & technology and technology transfer.

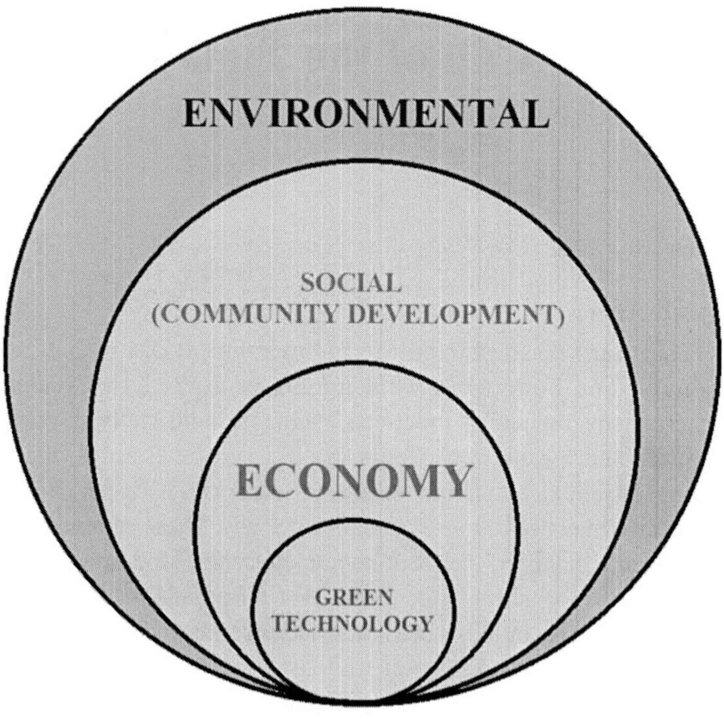

Figure 1. Four Pillars of Sustainable Development

This year's topic for the World Environment Day is 'Eat, Save – Reduce your Food Print'. There is huge wastage of food in urban life style. Massive awareness should be created among the people living in cities and towns to prevent food wastage, to adopt sustainable food habits etc. More incentives are to be given for opting birth control options so that population growth is brought down to less than 1% at the earliest.

Our modern way of living is based on unsustainable consumerism. The social repercussions of the unbalanced use of global resources are clear: social inequality, unemployment, emigration, poverty and harmful effects on our health. With our current patterns of production and consumption, we are rapidly approaching the natural and social absorption threshold limits. These developments should be counteracted by SD.The energy source from fossil fuels puts tremendous adverse pressure on environment, and our activity reduces biodiversity, which provides us food, shelter material, raw material for medicines, gases for survival, ecosystem services like recycling of water and other materials. The concept SD developed as the present model of unlimited growth in a limited environment is not the right answer to the complex of the problem that every country faces.

The Rio+20—2012 gave emphasis on Green Economy, which includes all aspects of SD but emphasizes on the adoption of Green Technology. The essence is using of green technologies in manufacturing and other sectors so that the GHG load to the environment will be substantially reduced to lessen the effect of climate change and to increase productivity by adopting green technology.

The world population is rising and predicted to reach 9 billion by 2050.Most of this growth will take place in developing countries. With huge population increase, there will be enormous pressure on the resources and to provide food and water to people will become a daunting task for the governments. India accounts for the largest proportion of malnourished children in the world. The world will face several challenges in health care, food and energy security, and fresh water availability. India exhibits very high variability; on one hand we have high GDP growth rates, a big pool of scientific manpower, a large middle class, malls, global retail chains, expanding multi-millionaires, and on the other hand huge number of people under poverty, unemployment and debt, with limited access to health facility and education, and most of them live rural areas and slum areas of urban centres. High level consumerism is seen especially among urban living people and they also put huge pressure on food, energy, water and many other resources.

Millennium Development Goals: In September 2000, world leaders came together at United Nations Headquarters in New York to adopt the United Nations Millennium Declaration, committing their respective nations to a new global partnership to reduce extreme poverty and setting out a series of time-bound targets - with a deadline of 2015 - that have become known as the Millennium Development Goals. Eight issues were included. These are (i) to end poverty and hunger ; (ii) to provide Universal Education; (iii) to promote Gender Equality; (iv) to care for Child Health; and (v) Maternal health; (vi) to Combat HIV/AIDS; (vii) to promote Environmental Sustainability; and (viii) Global Partnership. This aspect of human development are also intrinsically linked to SD end poverty and hunger in India and in the world, is linked to individual and societal food print.

Human Population Growth

The world population is rising (Figure 2, &Table 1), and predicted to reach 9 billion by 2050. Most of the population increase will take place in developing countries and there will be enormous pressure on the resources. The world will face several challenges in health care, food and energy availability etc (Dash, 2012). The demographics of India indicate that it is the second most populous country in the world, with over 1.21 billion people (Census2011), more than a sixth of the world's population. Already containing 17.5% of the world's population, India is projected to be the world's most populous country by 2025, surpassing China, its population exceeding 1.6 billion people by 2050. Its population growth rate is 1.41%, ranking 93rd in the world. India has more than 50% of its population below the age of 25 years and more than 65% are below the age of 35 years. It is expected that, in 2020, the average age of an Indian will be 29 years, compared to 37 for China and 48 for Japan; and, by 2030 (Census, 2011). The situation has started creating huge socio-ecological problem, which is to be addressed properly if we like to survive as a nation with dignity. Food security is important.

India accounts for the largest proportion of malnourished children in the world because of poverty and no availability of food, especially balanced food. India exhibits very high variability; on one hand we have high GDP growth rates, a big pool of scientific manpower, a large middle class, malls, global retail chains, expanding multi-millionaires, and on the other hand huge number of people under poverty, (some people also search public dust bins for left-over food), unemployment and debt, with limited access to health facility and education..

Table 1.World Human Population Growth since 10, 000 B.C. (based on Bhende, 2005, www.google.com/population/en.wikipedia.org)

B.C / A.D.	Population (estimated & census data)	Growth in % Doubling	Time (Years)
10, 000 B.C.	5 million	negligible	
5000 B.C.	10 million	negligible	
1 AD	250 million	<0.1	many centuries
1000 AD	350 million	<0.1	
1500 AD	450 million	0.1	
1600 AD	500 million	0.1	1600
1700 AD	600 million	0.2	
1800 AD	1000 million	0.3	200
1900 AD	1500 million	0.5	
1930 AD	2000 million	1.0	130
1960 AD	3000 million	2.0	
1975 AD	4000 million	1.8	45
1987 AD	5000 million	1.7	
1999 AD	6000 million	1.5	
2011AD	7000 million (7 billion) (India: >1210 million, Census 2011)	1.3 (India 1.41)	
2030AD	8000 million (8 billion)		55 (estimated
2050	9000 million (Estimated) (>1600 million-Estimated for India)		

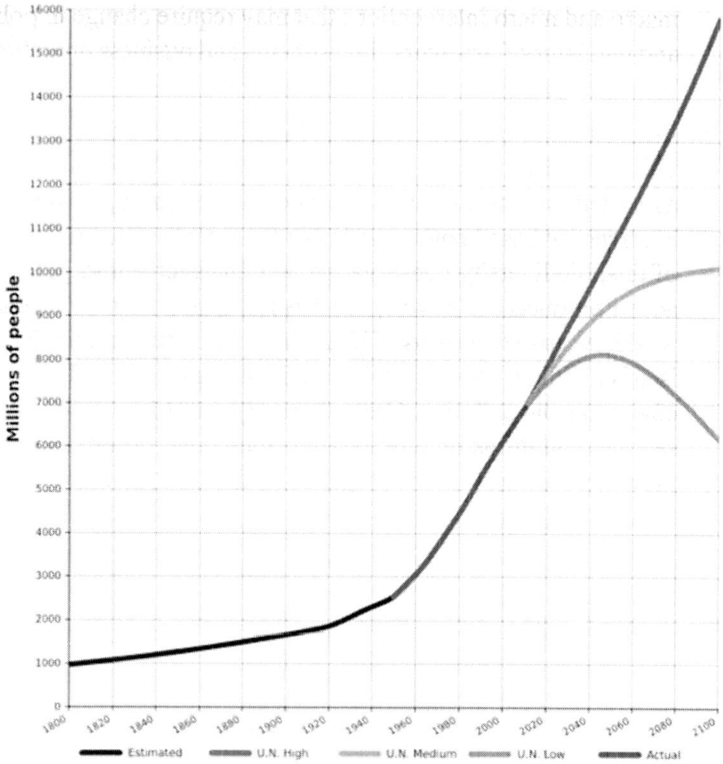

Figure 2. Human population Growth in the World

To make provision for food, water, energy, houses etc is a daunting task for any government or society unless every aspect is planned and executed properly. Population control and increasing

Food production using green technology, changing mind set for the consumption pattern, reducing waste, and following a sustainable way are the call of the time.

The prospects of Green Economy can be summarized as follows:

1. Protecting and conserving ecosystems and biodiversity,
2. Strengthening the three pillars such as (i) Economic development, (ii) Social development & (ii) Promoting environmental protection,
3. Discouraging unsustainable consumption & practicing sustainable consumption. Green economy, productivity & sustainable consumption are mutually supportive and may require

macro and micro interventions that may require change in policy and regulatory framework, investments and business operations. This may induce behavioral change in the society,

4. Eradication of poverty- About 70% of India's population live in villages and about 50% of them depends upon biodiversity such as agriculture, forest products, animal husbandry, fishing and biodiversity based cottage industry for their livelihood. In view of this, biodiversity conservation and management to get value added products for poverty eradication is important. Economic wealth measured in terms of High Gross Domestic Product (GDP) is often linked to overexploitation of resources and causing environmental pollution. Green economy aims to increase to basic services and infrastructure for alleviation poverty and helping in the overall improvement of quality of life.
5. Green economy & Green Technology should generate employment. New avenues such as renewable energy, new building retrofits for energy efficiency, public transportation, organic agriculture and vermi-technology, reclamation of degraded lands and forests, recycling and reuse of waste materials etc. These practices are expected to provide economic benefit, employment benefit and environmental benefits.
6. Financial Incentive:. The green economy policies may involve public-private partnerships as an economic instrument. This may involve public finance and fiscal measures, utilization of public expenditure on productive research and development, inclusion of environmental costs on goods and services, promoting purchange of green products and services, putting levies on pollution, and providing incentives on tax amount to industries for adopting green technology, strengthening legal framework. Governments can adopt appropriate policies depending upon the capability of the country to promote green investment
7. Benefits of Green Economy: It will generate clean environment by minimizing waste generation, reuse, recycling etc, provide security for getting clean water, renewable energy (energy security), change manufacturing philosophy, quality of transportation, and food security etc. For example about 70, 000 tons per year of plastic products are recycled in USA generating about 69 million dollars. In USA recycling of about 6.1 million

tons of urban waste in five years has generated more than 1. 6 billion dollars. Recycling of wastes instead of land filling or incinerating reduced about 412, 000 tons of carbon dioxide emission (source: Paribesh Samachar, 2012).

With the background of Rio-1992 to Rio+20-2012, I wish to discuss mainly two issues such as (i) Energy use, energy security and climate change and where do us stand ?; (ii) Food and nutritional security in the context of Indian situation.

Energy Source and Use

Energy is the key for survival, continuance and for sustainable development. There is increased need of energy in the agriculture, industrial and service sectors. Modern way and standard of living demand huge energy from unsustainable consumerism, transport facilities and increased construction work and other aspects of human use. Excessive consumerism is a factor for global warming and resource depletion. Patterns of consumption, food habits and recycling processes have influenced growth in energy demand and green house gas (GHG) emissions. The specific GHG emissions from food production and processing are much lower in developing countries including India than in developed countries. Besides, the consumption pattern is many industrially developed countries is not sustainable and put pressure on the society at large. (Figure 2).On priority basis industrially developed countries must take action to change the consumption pattern in their society.

Solar energy is abundantly available but we have not been able to trap it as per our need because of many reasons. The recent thrust is to utilize the renewable to address the GHG load in the atmosphere and environmental degradation.

Food Security

In 2011 the global population reached 7 billion. India's population reached 1.21 billion with annual growth of 1.41 per cent. The global population has been estimated to reach 9 billion by 2050 and India's population will reach 1.6 billion. It is obvious that there will be huge demand on food. The 'Green Revolution' of 1960s helped India to overcome chronic shortage of food and reduced hunger. Based on FAO report, a recent report of Government of India and Association of Biotechnologist led Enterprises (SPDash, 2012) has revealed that the number In of people who do not have food security increased from 800 million in 1998 to 1 billion in 2009 in the world because of rise of food prices from 2006 to 2008.

India about 25% people are classified as Below Poverty Line (Census, 2011). It means about 360 million people do not have food security.

Dash (2012) report quoting FAO report mentions that in recent times the dietary preferences change in developing countries and because of population rise and changed lifestyle, there will be 50% rise in cereals and 85% rise in demand of meat products between 2000 and 2030 and the food demand will double by 2050 putting enormous pressure on land, water and energy requirements. Indian census (2011) finds that about 69% of India's population (about 800 million) lives in villages. About 40-50% of them depend upon biodiversity such as agriculture, forest products, animal husbandry, fishing and biodiversity based cottage industry for their sustained livelihood and providing them food. In view of this, biodiversity conservation and management to get value added products is important for poverty eradication (Dash, 2010).

The livelihood options and rural economy are largely bio-resource based, especially on agriculture & horticulture (crops & fruits) as farmers, marginal farmers, labourers etc, home garden raising vegetables, fruits, flower and selling in the local market; keeping local breed cows and buffaloes for milk, dung, ploughing; catching fish, prawn, turtle, from local river, rivulets, common property ponds, own ponds, Goatery for meat, milk, skin, small Poultry for meat and egg, keeping Pigs for meat, and skin, use local forests/ Gramya Jungle for firewood, fruits, honey, ayurved, plucking kendu leaf for making bidi, a type of cigarette; artisan work based on bamboo, cane, wood and other bio-resources. Traditional ecological knowledge and bio- resource management is important.

Food grain production in India increased from 95 million tons in 1967 and 108 million tonnes in 1970s to 234 million tonnes in 2008, and to about 250 million tons in 2011 (Prasad, 2009, TOI, 2012).On the average 207 kg of food grain was available per capita in 2011 in India and this amounts to on the average 567 gram of food grains per capita per day. Total crop production and yield has increased considerably and has met food requirement of the people but considering the fact India's population is growing @1.41 %, crop production has to be doubled for future requirement. Another green technology based green revolution is needed.

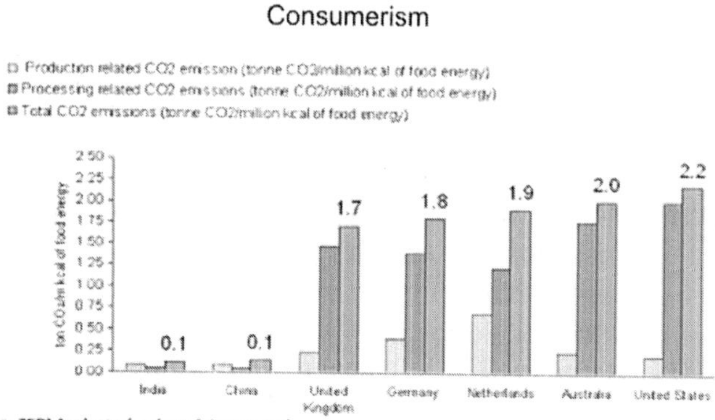

Source: TERI Analyses (various data sources)

Figure 3: CO_2 Emission from the food sector from field (Production) to table (processed food), excluding cooking

However the yield t/per ha varied from 1 in 1967 to 2.13 in 2011 for rice, 1.1 in 1967 to 2.9 in 2011 for wheat, 1.1 in 1967 to 1.9 in 2011 for maize, 0.5 to 0.6 for pulses (Prasad, 2009, en.wikipedia.org/ wiki- Agriculture_in_India). The arable land area has increased from about 120-122 million ha to about 159.7 million ha and the irrigated crop area is about 82.6 million ha. At present agriculture account for 16% of GDP and 10% of export earnings (Ayyapan, 2012 Orissa University of Agriculture and Technology convocation address).

Agriculture continues to be the largest sector in terms of employment, largely to the unskilled people as labourers. However the contribution of agriculture sector to India's GDP has fallen from 30% in 1990-1991 to 14.5% in 2011-2012 (SPDash 2012) indicating that the economy is changing from agrarian to other sectors. Agriculture remains for sustenance and livelihood largely for the rural people. The income from these occupations is subsistence level and does not meet the family requirements fully. To meet the challenge, agricultural productivity has to be increased and food should be available at affordable prices in India.

The option of converting more forest area for agriculture is not available as there is serious constraint of forest land area (Figure 4). Further conversion of forest area to agricultural land will create ecological backlash. Only solution is to increase the crop production through technology application and switch over to those Genetically Modified (GM) foods, which have stood the rigorous scientific testing and protocol and declared as safe for human consumption. Safety is the main concern of GM foods. Creating massive awareness among people to avoid wastage of food and storing of food should be done by the press and media and NGOs can play important role.

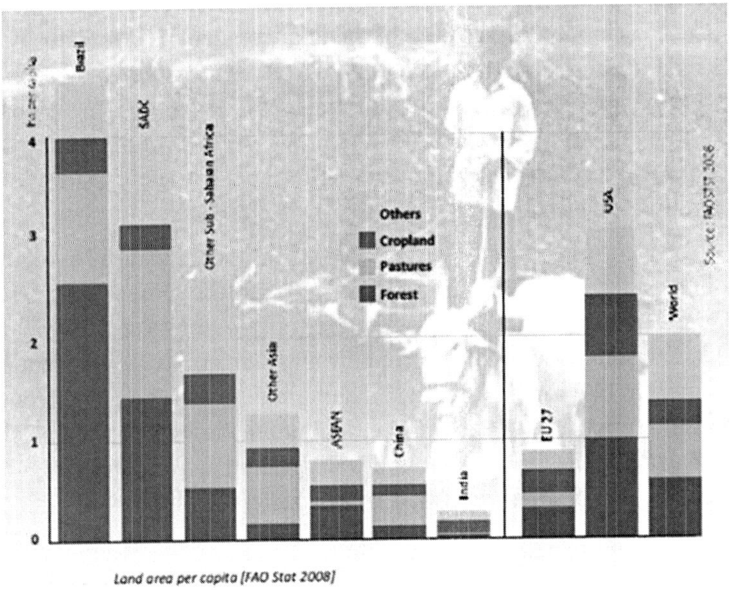

Figure 4. World Forest Cover as percentage of land area (source Dash, 2012, FAO 2008)

Considering India's forest cover and land area, it will not be sustainable to convert forest land to crop land. Only option is left to apply green technology and adopt population control measures.Biotechnology enabled the global increase in corn and soybean production by 130 million and 83 million tonnes respectively (Inderscience publishers 2011). Dash (2012) points out that biotech can improve agricultural production by promoting the use of disease resistant crops, enhancing the flavour and nutritional value of food products and subsequently impact global health and economies, especially among developing countries.

Water Availability for Agriculture

Of the total water used in India, agriculture is the most dominant users of water resource. About 50 to 60% of water resource is used for irrigation (Patnaik, 2006). Both surface and ground water is used for agriculture. India receives about 1050 to 1170 mm of precipitation annually. This is a huge water resource but almost 90% of this precipitation falls between mid-June and mid-October. Fourteen major river systems account for 85%of surface flow and share 83% of the drainage basin. They serve 80% of the total population. There are about 100 medium and minor rivers. The storage capacity is only 3.65 million m3. Of the total annual precipitation, India utilizes only 10-15%, which may increase to about 26% by 2025. We need to apply improved technology to increase water harvesting and storage capacity and ground water utilization.

The Central Ground Water Board has estimated that the available ground water in India is 210 billion m3 and the annual utilization potential is about 42.3 million hectare, only one-fourth of which is used at present. However, the ground water availability is not adequate in provinces like Tamil Nadu and Andhra Pradesh and excessive ground water use (unsustainable use) for agriculture in Punjab, Haryana, Delhi and Rajasthan has depleted ground water resource. More than 26 cubic miles of ground water vanished from aquifers in these states due to excessive agriculture use and may bring collapse in agriculture production unless remedial measures taken on priority basis (NASA satellite data-Yahoo internet news dated 14 August 2009).

Adequate ground water recharge and shifting to non water intensive crops are important to address the problem. Due to large scale deforestation and monsoon failure, there is a regular occurrence of drought in Kalahandi district and some other districts of Orissa, in Jharkhand, Bihar, Karnataka, Rajasthan, and Maharashtra. The supply of drinking water in Indian villages is not adequate, especially in summer. There is no organized water supply in most of the villages and even in some small towns.

Although there is good precipitation in India, sufficient care is to be taken to manage this water efficiently. The climate change has brought irregularity in incidence of monsoon rains and amount of rain fall has also become erratic. In view of this water storage facility should be created to harvest more water. Diversification crops are also required. Crops requiring less water for production should be chosen.

Soil Metagenomics

Soil Metagenomics (Figure 4) involves the analysis of a mixture of microbial genomes isolated from a soil sample in a particular ecosystem/habitat. Soil Metagenomics and it's interlink with soil fauna and soil fertility can lead to the mapping of soil metagenome to crop productivity. In this process the crop productivity can be increased by scientifically manipulating the soil metagenome without extensive utilization of pesticides and chemical fertilizers. Metagenomics studies can also improve the nutritional benefits of food crops; which will have a positive impact on public health (Figure 5, Dash 2012).

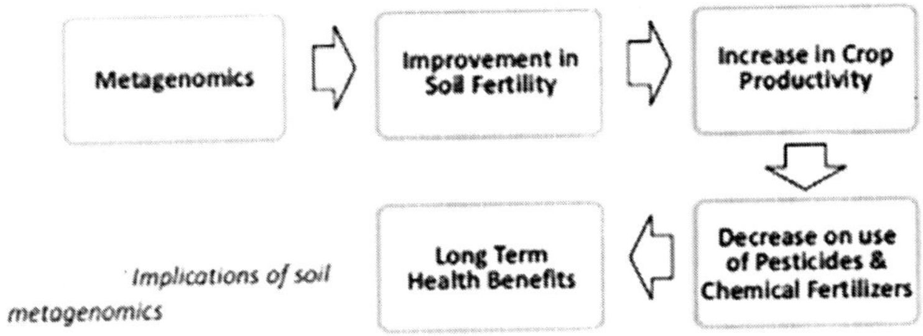

Figure 5. Implications of soil Metagenomics (source, Dash 2012)

Soil faunal biodiversity has important role in many land uses and for crop productivity (Dash et al, 2009).Studies on Soil biodiversity carried out in a project funded by UNEP-TSBF, Nairobi in multiple centres in India coordinated by Jawaharlal Nehru University from 2002-2010 have provided insight to the role of soil biodiversity, especially earthworms and soil fauna in variety of land uses. ICAR National Fund for Basic, Strategic and Frontier Application Research in Agriculture had initiated to study the dynamics of soil Metagenomics with participation of IARI, GB Pant University of Agriculture and Technology, and Tamilnadu Agricultural University (Dash 2012).M S Swaminathan Research Foundation has developed a biodiversity conservation model called the '4C Model' that takes care of conservation, cultivation, consumption and commerce requirements for crop plants. This model involves Gene-Seed-Grain Banks that help scientific selection and preservation of good seeds from land races (Dash 2012).

Soil Metagenomics and soil faunal studies in situ and ex situ should thrust areas of research so that the findings can be applied for increasing crop productivity to meet the challenges of the future.

Animal Protein

Fish, prawn, and turtle form part of the livelihood support in Indian villages. The total fish production (capture and culture) in India was about 8.29 million tons in 2010-2011, (10.17 million tonnes in 2015-2016, share of inland fishery rom fresh water and brackish water 6.57 million tonnes in 2015-2016) and out of this, fresh water aquaculture amounted to about 3.84 million tons, mariculture and coastal aquaculture 0.15 million tons and the capture fishery (both marine and fresh water) amounted to 4.30 million tons (Dilip Kumar, 2012).About 90% of fish production is consumed in the country. However the average per capita fish consumption is only 5.48kg compared to that of 12 kg internationally. However about 30-40 % people are vegetarian in India. In view of this the per capita fish consumption is estimated to be around 9kg among the non-vegetarian people.

There is good scope to increase capture fishery considering 8, 118 km of coastline and 0.506 million sq.km of continental shelf and 2.03 million sq.km of Exclusive Economic Zone (upto 22m depth). The area under ponds, tanks, beels and fresh water lakes etc amounts to about 3.42 million hectare. The current average aquaculture production is 600 kg/ha/yr & this yield can be increased considerably with innovative technology & management input. There is need of renovating village ponds for water harvesting and fish culture on priority. Rice-fish integration systems, cage culture etc are innovative systems, which should be widely practiced. The length of rivers and canals is 1, 71, 334 km. The cleaning of rivers and canals for fish culture will boost the village economy. Serious efforts are to be made using modern technology to increase the production to maintain at least 10-12kg of per capita consumption even if the population grows to 1.6 billion by 2050.

The animals (Table 2) are important for providing protein rich meat, milk and organic manure for agriculture, dung-fuel for rural homes, and by-products like bone meal etc as manures etc. However with reduction in grazing land and monetary constrains for stall feeding by the small and marginal land owners, government subsidy helps the farmers and the rural economy. But this is not the permanent solution. India has >483 million livestock; about 187million cattle, 97million buffaloes, 62million sheep, 120million goats, 14million pigs, 1million horses & mule, 1million camels, and 1million other livestock (Table 2). These animals provide meat, milk and produce a huge quantity of dung, which is utilized for biogas production. Besides 440 million poultry chicken are also available.

Table 2. Animal Resource of India (GOI, 2005, and Khillare et al 2012))

Species No of Breeds	Population 1997 (million)	Population2003 (million)
Cattle 30	19888	187.38
Buffalo 10	89.92	96.62
Sheep 40	57.49	61.79
Goat 20	122.72	120.10
Pig 3	13.29	14.14
Donkey	0.88	0.67
Horse 6	0.83	0.79
Mule	0.22	0.31
Camel 9	0.91	0.64
Yak	0.06	0.07
Mithun -	0.18	0.28
Rabbit -	0.40	
Poultry: Chicken 18	347.61	440.70
Duck 05		

Egg is the source of micronutrients. In view of this in the western countries egg is almost eaten every day and Egg is provided to the children. They get protein, and micronutrients which help in the brain and muscle development.

Religion and beliefs play important role in choosing type of food .Creation of awareness among people to choose protein and nutrients and a balance diet is important.

Indian leather industry includes 125 medium and large scale units, 1200 small scale units and thousands of tiny tanneries in rural areas. Hides and skins are terminal markets. The states of Tamilnadu, Uttar

Pradesh and West Bengal account for more than 80% of country's leather output.

Animal resource forms an important part in food security, especially for protein security and livelihood options for weaker section of the society. Fish and livestock yield can improve over all food security in India and income for rural people. Women are largely form the work force in animal farms and hence this sector helps in women empowerment through financial security. Marginal and small land owners are the owners of 71% of cattle, 83% of buffaloes, 88% of small ruminants, 70% of pigs and 74% of poultry (government of India-publication, 2005). In view of this the village livelihood sustainability is dependent on animal resources. The live-stock contributes about 25% of the agro-sector of the country in terms of money. This is the important sector for scientific research to increase the milk, egg and meat production. There is good scope to enhance this production and other welfare measures are required. Genetic selection of animal breeds, feed and nutrition aspects, health and disease aspects and opening of veterinary dispensaries and hospitals in villages are the priority areas.

Environmental Impact Assessment (EIA)

Since 1994 EIA study has become mandatory for all new projects or expansion or modernization of existing projects which have substantial stake on environment. As per 14 September 2006 notification of Ministry of Environment and Forests of Government of India, and subsequent amendments, the generic structure of environmental impact assessment document and check list of environmental impacts have been stipulated. The purpose is to assess the expected impacts and to address them through an environmental management plan if the developmental project is allowed to operate. The check list includes land environment, water environment, vegetation, fauna, air environment, aesthetics, and socio-economic aspects, building materials, energy conservation and environmental management plan. The check list is a reflection of livelihood options prevailing in rural India. However there is a big gap between the EIA stipulations and implementation. The sustainability concerns of Indian rural systems have not been adequately addressed. The interest of multinational companies and industrial houses of the country have received more attention than the natural resource management or furtherance of traditional livelihood options conserving the 'traditional ecological knowledge' available in village systems. Ramakrishna (2009) points out emphatically that 'very traditional societies living close to nature and natural resources around them may have to have traditional ecological knowledge being brought in to a much larger degree so as to avoid social disruptions setting in, compared to the more modern societies who may need traditional ecological knowledge only to be brought in so as to create buffering mechanisms within the socio-ecological system and thus cope up with the ill-effects arising from excessive use of energy-intensive technologies.'

My understanding is that sustainable development in Indian rural systems depends on creating facilities to enhance the existing livelihood options, adopting effective community participatory approaches, and sustainable management of bio-resources. However, this does not prevent opening of new avenues which will have minimum environmental impacts and can be addressed by application of green technology and will create livelihood options for youth. This is the time to strengthen the EIA protocol and its judicious implementation to address environmental uncertainties and to conserve bio-resources based traditional ecological knowledge and rural livelihood options of about 360 million people in the country.

References

Ayyapan, S. 2012. Convocation Address, Orissa University of Agriculture and Technology, Bhubaneswar.

Bhende, A. 2005. Principles of Population Studies, 5th edition, Himalayan Publishing House, Mumbai.

Brundtland, Gro Harlem. (Chairman of the Commission), 1987. 'Our Common Future', Oxford University Press, U.K.

Census, 2011. Government of India, New Delhi.

Dash, M. C. 2009 .Environmental Status and Sustainability Report As per Orissa Resettlement and Rehabilitation Policy, 2006, Prepared for ARDCOS, Bhubaneswar (Lanco-Babandh Power Private Limited, Denkanal, Orissa), 73 pages.

Dash, M.C .2009. Bioresources as a tool for food security and Sustainable Development

For Rural Livelihood in India in the context of Industrial Development and Environmental Protection: An Overview. The Ecoscan, 3 & 4:201-208.

Dash, M. C., Saxena, K. G. and Giri, S. 2009. Vermitechnology for Wasteland Reclamation, Plant Productivity and Composting: A Review in Indian Context. *International J. Ecology and Environmental Sciences.* 35 (2-3) :163-185.

Dash, M.C.2010. Environment, Energy and Development from Stockholm to Copenhagen and Beyond the Celebrations. The Bioscan: Special Issue, 1:1-11.

Dash, M.C.2012. Sustainable Development: The Journey from Rio (1992) to Rio+20 (2012) and Beyond. International Journal of Ecology and Environmental Science, 39 (1), 1-21.

Dash, Satya Prakash.2012 .Indian Biotechnology: The Roadmap to the next Decade and Beyond, Government of India, Department of Biotechnology and Association of Biotechnology Led Enterprises, 135 pages.

Dilip Kumar, 2012.Technology Driven Fisheries and Aquaculture for Rural Livelihoods, In: Sustainable Development of Fisheries and Livestock for Food Security (Khillare, Y.K., M.C.Dash, and B.N.Pandey, Ed, Narendra Publishing House, New Delhi. 41-52pages.

Food and Agriculture Organisation. Rome, 2008.

Forest Survey of India.2005, and 2009, 2011.Report, Forest survey of India, Dehradun.

Government of Orissa.2007.Integrated Social and Environmental Assessment Study, Prepared by Varun Techno Infrastructure Pvt. Ltd, World Bank Project) 2007. Volume 1 and 2.

Government of India Country Report on Animal Genetic Resources of India. 2005. Department of Animal Husbandry and Dairying, Ministry of Agriculture, Government of India, New Delhi.

http://en.wikipedia.org/wiki/Agriculture_in_India.

http://en.wikipedia.org/wiki/List_of Power_Stations_in_India.

http:google.com/population

India State of Forest Report., 2005, 2009, 2011.Summary, Forest Survey of India, Ministry of Environment and Forests, Government of India, New Delhi.

Khillare, Y.K., Dash, M.C. and B.N.Pandey.2012.Sustainable Development of Fisheries and Livestock for Food Security, Narendra Publishing House, New Delhi, 222 pages.

Patro, P.S.N.2012. We are in catastrophic Climate, Let us think for it. Managing Water Resources Towards a Green Economy, Indian Public Health Engineers (I), Regional Centre, Bhubaneswar, 29-36.

Pattanaik, S. 2006. Water Resources, In: State of Environment-Orissa (M.C.Dash, ed.), Orissa State Pollution Control Board, Ministry of Environment and Forests, Bhubaneswar.

Prasad, Rajendra. 2009. Food Production in India-Progress and Future Strategies, Proceedings of National Academy of Sciences, India, Section B. 79, 104-109.

Prayas Energy Group.2009. Prayas Group, Mumbai-Report.

Ramakrishnan, P.S. 2009. Towards Addressing Societal Concerns: Moving Through Genecology and Ecosystems to Socio-Ecological Systems, International Journal of Ecology and Environmental Sciences 35 (1) :13-33.

Times of India.2010, 2011 and 2012 Daily News Paper, Bhubaneswar Edition.

Teri (Tata Energy Research Institute), New Delhi.2009.Consumerism Analysis, 2009. (Various sources)

World Wide Fund for Nature.2009 and www.worldwatch.org

11

Environment, Energy, and Development from Stockholm to Copenhagen and Beyond the Celebrations

Introduction

Thirty eight years before on 5th June 1972 at Stockholm, the world realized the future of mankind is bleak unless we take urgent actions to protect our environment from the colossal degradation and damage we have inflicted on the planet in the last few hundred years and pledged to bring world wide awareness among the people to protect the environment, the mother Earth for our survival. The Stockholm conference lead to the establishment of UN Environment Programme and gave emphasis on protection of environment for the future. The World Commission on Environment and Development was established in 1983 and brought out its report 'Our Common Future' in 1987.This report highlighted on poverty, depletion of natural resources faster than exploitation, deterioration of environmental quality and quality of life. These actions lead to the organization of the Earth Summit in Rio de Janeiro in Brazil in June, 1992.

The Rio conference gave widespread visibility of the important concept of sustainable development. This concept placed environmental management in a position to address the environmental impacts of developmental projects so that the environment is protected for the present and future generations. The conference in Johannesburg, South Africa in 2002 provided opportunity to review the actions taken and benefits made and to review the different aspects of resource utilization and conservation. Biodiversity conservation received top priority. Besides, the important issue of energy use and climate change and its effect on developmental projects, on human activities and future of mankind was very much focused.

The theme in the World Environment Day on the 5th. Of June 2010 was "Many Species, One Planet, and One Eartth. The theme respectively in 2009 and 2008 was "The Earth needs us, UNite to combat Global Warming and save the Earth" and 'Carbon dioxide: Kick the habit for low carbon economy'. The UNO has declared 2010 as the 'International Year of Biodiversity Conservation'. The UNO has also declared 2011 as the 'International year for Forest Conservation'.This shows the growing importance of protecting the environment, conserving biodiversity, especially forests to combat climate change. Combating climate change involves the phasing out of the use of fossil source of energy and switching over to other forms of energy which do not carbonize the world. (Dash, 2010 a and b).

These annual sermons, however regarding the state of environment of earth clearly show that our modern way of living is at stake because the present style of progress and mode of development with unsustainable consumerism and energy source from fossil fuels eats away into our basic life supporting system and puts tremendous adverse pressure on environment, and reduces biodiversity, which provides us food, shelter material, raw material for medicines, gases for survival, other ecosystem services like recycling of water and other materials – something like legendary Indian poet Kalidasa, before being blessed by the goddess of learning was cutting the tree branch sitting over it.

Virtually all earth's ecosystems have been significantly transformed through human activities and the changes have been more rapid during last hundred years as the population grew faster and migration to urban centres became a necessity due to lop-sided development and management. It is now realized that the present concept of unlimited growth in a limited resource based environment is flawed and sustainable development is necessary.

With huge population increase, especially in underdeveloped and developing countries and new types of ailment looming on the horizon consequent to a synthetic life emptying the natural medicine tool box consequent to extinction of the ecoservers (medicinal plants) is disastrous. Who knows what types of metabolic assaults humanity will face in future in Nature's feed-back loop? In view of this the biodiversity conservation is very important.

Energy Use

There is increased need of energy in the farm, industrial and service sectors. Modern way and standard of living demand huge energy from unsustainable consumerism, transport facilities and increased construction work and other aspects of human use. Energy is the key for survival, continuance and for sustainable development. Excessive consumerism is a factor for global warming. Patterns of consumption, food habits and recycling processes have influenced growth in energy demand and GHG emissions. The specific GHG emissions from food production and processing are much lower in India and other developing countries than in developed countries.

Solar energy is abundantly available but we have not been able to trap it as per our need because of many reasons. The recent thrust is to utilize the renewable to address the environmental dedradation.

India's Energy Source

About 70% of our energy source (1, 53, 694.09 MW) is fossil fuels, especially coal and lignite. The distribution is as follows:

- 81, 355.88 MW-coal based,
- 16, 822.85 MW –Gas based
- 1199.75 MW diesel based.

Thermal power plants are not environment friendly as these plants generate huge amount of green house gases and pollutants such as SO2, NOx, Cox,, SPM, Fly & bottom ash, metallic dust, etc etc. About 120-150 million tons of fly ash is generated annually in India (Singh, 2009 and Jain, 2010). The utilization is only about 25-30% in brick making, road construction, forest soil fertilization, and in agriculture etc (Jain, 2010). Hence it is a huge problem. One estimate (Paribesh Samachar,

2010) shows that the pollution load from 1 MW Indian coal based thermal power generation amounts to generation of:

- ~19 tons of CO_2 per day .MW
- >136 kg of SO2 per day.MW
- >7 tons of fly ash per day.MW
- Particulate matter 60kg per day.MW

Since about ~1, 53, 694 MW electricity is generated the pollution load per day on the environment is huge i.e. 2.92 million tons of CO_2, 20, 902 tons of SO2, 1 million tons of flashy, and 9222 tons of particulate matter. The pollution potential and the Green House Gas generation by the thermal power plants are huge.

In view of this, there is now a policy shift for energy generation by government of India. More stress is being given to renewable source of energy, especially solar, wind, biomass gasification and nuclear in 11th plan and onwards for reduction of green house gas emissions, pollution and for conservation of resources (Prime Minister's eight missions including the solar and energy efficiency missions, 2009). Indigenous technology development and appropriate technology transfer mechanism, collaboration of public sector-private sector in all sectors i.e. agriculture, industry, animal husbandry and fisheries, forestry, water resource management, energy generation, environment etc are required in the present juncture of this transition.

In 1947 India had electricity generating capacity of only 1470 MW with about 49% load factor .At present India has the capacity to generate >1, 53, 694.09 MW (105 times growth) with about 80% load factor. In spite of this increase, India is not able to meet the demand of energy from the farm, industry and service sectors. Hence the situation has become critical. In the 10th and 11th plans, additional 27, 283 MW & 78, 700 MW could be achieved. It has been estimated that by 2016, the energy production capacity will reach 2, 15, 804 MW (Fatesingh, 2010).

In spite of this huge expenditure, per capita energy consumption in India is only about 500 units per year compared to about 2600 units at international level in industrially developed countries. India has 618 districts, and 5, 93, 732 villages. Out of this, about one lakh villages do not have electricity supply till today. (It is expected that during this plan all villages will have electric supply).Out 13, 82, 71, 559 households, only 7, 80, 90, 874 household have received electricity supply. In India 43.5% people are deprived of electrical energy, & 30 % people use <50 unit per year. (Dash, 2010)

Climate Change: Climate Control Policies (Kyoto Protocol 1997)

In 1997, the global community met in Kyoto in Japan to work out mechanism for controlling Green house (GHG) emissions. Policy decisions taken during this conference have been termed as *Kyoto Protocol*. According to the mandate of this Protocol, all countries were grouped into two categories. Those consuming energy above the global average mostly the industrially advanced countries like European countries, USA, Canada, Japan, Australia, and Russia, are called Annex-1 countries (Developed) and the rest are called Non-Annex-1 countries (Developing). According to Kyoto Protocol, Annex-1 countries have to bring down their GHG emissions to less by 5.2 % average from the base level of 1990 keeping in mind their economic development, population growth, etc. If for some reason they are not in a position to bring down the emission to below 1990 level, they are allowed to go and reduce the GHG emission in the developing countries where the efficiency of energy consumption is also lower and there is a need to consume more energy for ensuring their development. Any reduction made in the specific green house gas (GHG) emissions can be considered as being reduced in the respective country providing the monetary benefit.

Any organization or industry in the Non-Annex-1 country reducing the GHG emissions can trade those emissions with the Annex-1 countries. The Kyoto Protocol further stipulated that the time period allowed for GHG emission reduction commences from January 2000 and ends up by 2012. The accounting for Annex-1 countries begins from 2008 and ends up with 2012 (first crediting period). The process is controlled by the respective governments and the UNO.

To reduce the global warming potential (GWP) from the GHG, industries can undertake any of the following CDM projects: Increase in methane gas (biogas) consumption, Increasing the use of biomass in place of fossil fuel for energy generation, Fuel switching from coal to biomass (biomass gasification), Switching over to hydroelectric power, Energy efficiency projects, Fuel switching from coal to gas,

Improving energy efficiency at the user end, Improved Agriculture practices and to increase carbon sink potential.

This financial instrument of selling GHG emission is called carbon trade under Clean Development Mechanism. Kyoto protocol does not necessarily reduce industrial GHG emissions in the world taken in whole, although it aids to aforestation and forest conservation.

Carbon Trading essentially means exchanging points earned by the reduction of GHG for money. This has emerged as a huge opportunity for Developing countries where GHG emissions are much below the quota fixed. Developing Economics are not expected to meet emission targets during this period. Companies in Developing Countries can earn carbon credits by investing in Clean Technology. The principle is based on the fact that climate change is a global problem that requires a global solution. Global cooperation on technology transfer is important and developing countries need access to climate-friendly technology. Global Environmental Facility (GEF) has a critical role to play in the co-development and transfer of technology and it can be made through several different channels. Technology transfer must be accompanied by capacity building. Though UNFCCC has made substantial progress in evolving the mechanism for inter-state sale of carbon emission credits; there are still some gray areas, which are being resolved from time to time.

Carbon trading is in full swing in many developing countries like Brazil, Mexico, China and India despite the fact that USA and Australia did not ratify the Kyoto Protocol.

Carbon trading is moving across all the countries in a very rapid way and expected to involve 1000 billion dollars trading by 2012. The organizations can reduce energy consumption or release of any of the six GHG gases by adopting cleaner options (Clean Development Mechanism) and claim reduction in GHG emission. Renewable and methane captures are at the top of the list and energy efficiency at lower end. As the accounting period approaches (2012) the prices may increase further. Size of the project is determined based on the total emissions saleable over accrediting period.

Developed nations, especially USA, Japan and Australia wanted that large economies like China, India, Brazil, and Mexico should take some steps to control carbon emission although they are listed under Annex-2 countries. This was the scenario before 2007.

Bali Road Map (2007)

UNO organized the conference in 2007 at Bali. The slogan was 'NO TIME LEFT- ACT NOW'. There were 10000 Participants, 187 Govt. Representatives and the conference was held from December-3-14, 2007 and the conference adopted Bali Road Map: a new negotiating process was to be concluded in 2009 & will lead post 2012 and lead to international agreements. USA agreed for Bali Road Map of December, 2007. The salient Features were:

1. Lunching adoption fund (AF),
2. Reducing GHG emissions;
3. Technology Transfer is to made simple
4. USA & Australia agreed for post 2012 agreement & they argued for some commitment from developing countries, especially from China, Brazil and India for reduction of GHG.
5. The Message was 'The World can not afford to wait'.

Countries are to address the question of a shared vision of long term cooperative action-carbon emission reduction taking into account social and economic conditions and other relevant factors.

Countries must negotiate the question of 'Enhanced action on technology' but countries shrink when clarifying what they wish to do mitigate climate change.

Poznamn, Poland-Climate Change Conference (December 1-12, 2008) :

UNFCC-14th Conference of Parties (COP) to solve the problem of erosion of trust between developed and developing countries on implementation of Bali Road Map was held at Poznan.Over 3000 delegates from 186 countries participated in the conference. Developing countries need huge amount of money every year till 2015 to adapt to climate change effects according to the Human Development Report of UNDP, 2007.At present the availability of the fund is meager and developed countries are not contributing as expected.GEF has been handling money disbursals. Least developed countries (LDC) found the process too much cumbersome. Besides UNFCCC has found that CDM process can not properly account for energy efficiency projects in the building sector.

Policy Shift by USA

There was a policy shift in USA with regard to Climate Change after Mr. Barack Obama became President and Mrs. Hillary Rodham Clinton was appointed as Secretary of State. I quote some of the portions of the address of the Secretary of State Mrs. Clinton delivered at 17 major Economies forum on Energy, Climate on 17 April, 2009.

'It is an environmental issue, a health issue, an economic issue, an energy issue and a security issue. It is a threat that is global in scope, but also local and national in impact.' Desertification and rising sea level generate increased competition for food, water and resources. But we also have seen increasingly the dangers that these transpose to the stability of societies and governments. We see how this can breed conflict, unrest and forced migration. She stresses on four points. These are:

1. Climate change is a clear and present danger to our world and demands immediate attention, the science is conclusive.2. The US is determined to make up for lost time, both at home and abroad and the USA is committed to address this issue and to act., 3. The countries should pull together and work towards a successful out come of the UN climate negotiations later in the year at Copenhagen, 4. We all must cooperate and new policy and new technologies are needed to resolve this crisis. The USA has made budget provisions for funding and for providing loan for clean energy development, targeting to double supply of renewable energy.

The international business on CDM will be huge; their priority area is to make the homes and buildings more energy efficient.

On one hand it is very encouraging but there is an element of business interest which makes developing countries worried. They feared that the developed countries may vouch for imposing restrictions on carbon emissions. (Times of India dated 23 May 2009.

Every nation wants its enhanced economic development, although shows concern for global warming and emissions.

Bonn, Germany Conference

The first draft for negotiations on the Climate change policy beyond 2012 released on 19th May 2009 at Bonn, Germany included two options for developing economies like India to deviate their emissions from the current base line by 15-30% by 2020 or reduce by 25% from 2000 level by 2050.This generated lot of heat in the negotiations. This provision deviates from UNFCCC's original text and also the spirit of Kyoto protocol.

The 20 nation Major Economic Forum (MEF) met in London on 19th & 20th October, 2009 and suggested that International monitoring of GHG should be mandatory and the proposal was not accepted by the developing countries.

Copenhagen: December 7-18, 2009

The Copenhagen conference was attended by 193 countries & 120 Heads of States. The deliberations resulted in developing the following:

- No specific emission targets set for the developed countries
- Consensus was to keep temperature rise to below two degrees Celsius by 2050, but no strong commitment was made.
- Emphasis was on mobilization of financial resources for supporting reforestation efforts of developing countries
- Scaled-up funding for poor countries, including long term cooperative action (fast-tracked fund) of $30 billion during 2010-2012.USA will take initiative to mobilize $100 billion by 2020

Summary of Copenhagen Conference- Proceedings:

- Most of Annex-1 countries wanted a new deal replacing Kyoto Protocol (5.2% cut of 1990 level). Non-Annex countries insisted of retaining Kyoto and action in 2008-2012.Kyoto Protocol survived-discussions to continue
- Consensus (120 out of 193 countries) agreed to allow up to 450 ppm CO_2 & 2oC rise in temp by 2050.
- Outcome was voluntary emission cuts by BASIC countries and retention of Kyoto. (A political agreement was signed- neither supported by UNO not a legal document)
- International monitoring was not accepted.
- Consultations yield a political accord not a legally binding treaty.
- UNO-UNFCCC meeting in Mexico in 2010 (November 29, 2010 to December 9, 2010) did not introduce punitive measures for violation of International agreements. (?).

- Rich nations to take on internationally-binding commitments on reducing GHG (Kyoto, Bali protocols) & Developing countries to take domestic action to cut the GHG emissions- no legal document.
- UNO to play more active role in the governance of flow of finances to help developing countries counter climate change
- Countries agreed to scale up public finances significantly for GHG reduction.

India's Voluntary Policy Shift

Considering 8% GDP growth rate per annum in 2002-2010, the estimate of GHG from all sources in 2000 in India was about 1485 million tons, per capita emissions about 1.1 tons... This was about 3% of total global emissions. An assessment of the current and projected trends indicate that green house gas emissions grew @ 4% annum during 1990 & 2000 and projected to grow further due to developmental needs. The absolute level will not be less than 5% of global emissions in 2020. (Sharma et al, 2006. Shukla, 2006, FSI, 2009).In India, the estimate is 2115 million tons of GHG emission in 2010 and >2800million tons of GHG emission by 2020. Indian government in Parliament has announced 25% cut in emission level by 2020 taking 2005 as the base year before the Copenhagen meet. It means that the dependence on energy from coal to renewable will receive priority.

Prime Minister's council on climate change has approved eight missions under India's National Action Plan on Climate Change.

- achieving energy efficiency; 'Perform, Achieve and Trade' (PAT) mechanism
- promoting Renewable energy;
- promoting sustainable development especially through clean technology development & transfer,
- promoting afforestation, & enhanced forestation (from 20.6% to 33% of land cover) to increase sink factor
- promoting sustainable agriculture
- The PM's Office has announced that by 2015, India will save about 5% of annual energy consumption & nearly 100 million tons of CO_2 emissions per year.

- The National Solar Mission has approved enhancement of solar energy generation.
- On 21st October 2009, India & China have signed a MOU for collaborating on climate change issues at mutual and international level
- India launched 83Kg small observation satellite (IMS-I) in April 2008
- On 18th October 2008, Mr.G.Madhavan Nair announced further actions to be taken: India plans to launch two climate satellites (One 50 Kg micro satellite for climate study and the Second satellite (Remote sensing) will monitor GHG (CO_2, & CH4) emission in polar orbit
- The Satellite will contain 2 optical instruments-
  - (i) a multispectral camera,
  - (ii) a hyper-spectral camera to operate in visible and near infrared regions.
- These actions indicate India's concern and positive voluntary actions.

Sustainable Development

Sustainable Dev elopement is the development which meets the needs of the present without compromising with the necessities of the future generations (Brundtland Commission, 1987).This concept developed as the present model of unlimited growth in a limited environment is not the right answer to the complex of the problem that every country face. This is indeed a tall order to achieve. It also means redistribution of resources as it will reduce inequalities. Redistribution may include technology transfer, financial aid and compensation to prevent environmental degradation, resource conservation etc.

How did the concept of sustainability come about?

In 14th century, "sustainable" forestry was practiced in India based on the principle of not cutting down more trees than the growth rate for replacement allowed. In India the concept is very old & linked to Vedic times.

In Europe, the concept of sustainability can be traced back to 18th century. In 1795, the economist George Ludwig wrote that the forests are to be used in such a way that "future generations will have just as many advantages as the present generation does". The concept of sustainable development was first introduced by the International Union for Conservation of Nature and Natural Resources (IUCN) in 1980. A definition of sustainable development was first given by the World Commission on Environment and Development (popularly known as Brundtland Commission) in 1987 as a development process that "meets the needs of the present without compromising the ability of the future generations to meet their own needs .The Brundtland Commission considered population control, food security and energy supply as critical components of sustainability. Since the Rio Summit in June 1992 and the adoption of Agenda 21 by the global community, many nations have set sustainability as a key goal of their development. According to the World Resources Institute, the four dimensions of sustainable development are social, economic, environmental and technological well being. These dimensions provide a framework within which one can specify details for countries at a given stage of development.

The environmental laws, the application of science-technology and creation of public awareness are defining factors for the success of sustainable development.

Concern over the state of the environment started building up since 1960s due to decline in environmental quality & deterioration in living conditions. UNO became concerned & it was reflected at the UNO-conference on the Human Environment held in Stockholm from 5th June to 12th June, 1972, Rio conference, 1992, Review of Progress in Johannesburg, 2002. Government of India actively participated in1972 conference and follow-up conferences and the then prime minister presented a thought provoking paper on poverty & environment and strongly voiced concerns in 1972 conference.

Several measures were taken in India for Environmental Protection before the Stockholm conference but there was no general legislation on a holistic manner to protect the environment. In India the existing laws dealt directly or indirectly with several environmental issues like forest conservation, wild life conservation etc but no general comprehensive law existed to deal with pollution prevention, control

& abatement. India has witnessed manifold increase in Industrial sector & the increasing tendency of industrialization & urbanization is growing very fast.

In 42nd amendment of the constitution, two articles were inserted and these have direct bearing on Environment:

Article 48A: The state shall Endeavour to protect and improve the environment and to safeguard the forests and wild life of the country.

Article 51 – (g) of the Constitution states that:-

'It shall be the duty of every citizen of India to protect and improve the natural environment including forest, lakes, rivers and wildlife and to have compassion for living creatures

As per Article 21, 'No person shall be deprived of his life or personal liberty except according to procedure established by law' The Apex court has interpreted this article to include the rights of a ' living environment congenial to human existence' while disposing the Vellore tannery case in 1996 and ordered for implementation of 'Polluter pays' principle. It shows the concern of the Hon'ble Supreme Court to Environmental Pollution and Suffering of the people.

Public Interest Litigation provision was introduced under article 39A (42nd amendment) (equal justice & free legal aid). As per Article 47, 'The state to ensure as its primary duty of (I) Raising standards of living of its people; (ii) To increase the level of nutrition of the people and to bring improvement in public health.

Science & Technology has also advanced and methods are available to control pollution and prevent environmental degradation.

The hindrance in proper implementation of laws in India is non availability of adequate public awareness, weak environmental governance and corruption. Legal matters on environment protection are not addressed on time, due to many reasons etc.

Max Muller, the celebrated German Scholar once wrote '*if we were to look over the whole world to find out the country most richly endowed with all the wealth, power and beauty which nature can bestow - in some parts a veritable paradise on earth – I should point to India*'.

What he wrote of India, in fact, is truer for the Rural India which is a unique repository of all those beautiful things that God and human hands created for the posterity to draw inspiration for a better tomorrow. India is a country of villages (618 districts & 5, 93, 732 villages- 2001 census)

The livelihood options and rural economy are largely bio-resource based, especially on

- Agriculture & horticulture (crops & fruits)
- home garden (vegetables, fruits, flower)
- Diary (milk, cow dung, ploughing)
- Fishery (fish, prawn, turtle)
- Goatery (meat, milk, skin)
- Poultry (meat, egg)
- Piggery (meat, skin)
- Forestry/ Gramya Jungle (firwood, fruits, honey, ayurved)

Artisan work based on bamboo, cane, wood and other bio-resources.

Traditional ecological knowledge and bio- resource management is important.

Recent thrust on industrialisation, especially in resource rich but economically poor states like Orissa, Jharkhand, Chhatisgarh, Madhya Pradesh, and some other states threaten the traditional livelihood options Therefore there is a need for implementation of EIA, EMP and Bio-safety protocol very strictly and judiciously. The forest provides firewood, house building material, fodder, fruits, medicinal plants and other utility materials to the villages with some control by village Panchayat. The runoff water from the slopes of the mountain forest/ Gramya jungle is stored in the water harvesting village ponds.

The villagers depend upon the buffaloes and cows for milk, goat and sheep, poultry for meat and eggs and use the bullocks and buffaloes in the crop field for ploughing. The goat rearing and poultry is a profitable business for many village households. The village forest and grasslands are also used extensively for grazing of domestic animals of the villages

In the changed situation of industrialization, the reduction/complete loss of grazing lands, villagers have problems of providing grazing land and unable to afford stall feeding for domestic animals.

However, employment opportunities for villagers in the changed scenario have increased their income, which help to partially stall feed the animals. The home gardens are also productive resources and are to be managed well. Ecosystem people (Dasmann, 1988, Gadgil and Heda, 2009) are largely forest and bioresource dependent. The livelihood pattern in villages varies according to cast, ethnicity and financial condition of the families. Since the livelihood pattern is largely bio-resource based, government schemes should protect these resources and funding should be done to enhance bio-resource productivity and management.

The income from these occupations is subsistence level and do not meet the family expenditure. In view of this migration to urban centers inside the states or to distant places in search of work to augment family income happens

Since 1994 EIA study has become mandatory for all new projects or expansion or modernization of existing projects which have substantial stake on environment. As per September 2006 notification of Ministry of Environment and Forests of Government of India, the generic structure of environmental impact assessment document and check list of environmental impacts have been stipulated. The purpose is to assess the expected impacts and to address them through an environmental management plan if the developmental project is allowed to operate. However there is a big gap between the EIA stipulations and implementation. The sustainability concerns of Indian villages are yet to be adequately addressed. The judicious implementation of EIA provisions is need of the time

The natural resource management or furtherance of traditional livelihood options conserving 'traditional ecological knowledge' available in village systems is to be focussed and managed well. Sustainable development in Indian villages should largely centre on creating facilities to enhance the existing livelihood options, adopting effective community participatory approaches, and sustainable management of bio-resources. This does not prevent opening of new avenues which should have minimum environmental impacts and can be addressed by application of science and technology. This will create additional livelihood options for educated youth.

The integration of Science-Technology, Traditional Ecological Knowledge, and Judicious application of EIA protocol can make significant contribution to protect bio-resources for rural livelihood options for sustainable development

For a baby, Mid-wife does not substitute a mother, nor do parks and gardens in city substitute natural forest. Even a tractor is considered to be a poor substitute environmentally to a pair of cattle on the plough, because, the tractor consumes fossil fuel brought from outside, generates pollution, needs repair and spare parts and unlike the cattle does not produce dung which goes into the socio-agricultural system of India. Cattle are a bioresource and have a life cycle to sustain the population. No wonder, since history began this bovine animal has been catalysis in stabilizing the Indian society in general and families in particular. Whereas, the tractor or any such mechanical equivalence of modern endeavor is found to breed a series of unrest in ecosensitive societies. I do not advocate replacing the tractor by a pair of cattle on the plough, but how much ecofriendly it would be to run the tractor not by fossil fuel but by renewable energy.

It is our sacred duty to protect the environment, the natural ecosystem services, conserve biodiversity handed over to us by the Mother Earth and practice sustainable development in all our activities including our day to day living is call of the hour.

References

A Report of the Intergovernmental Panel of Climate Change, 2007 (4 volumes), UNEP, Switzerland.

Centre for Science and Environment (CSE), New Delhi.2009. Climate Change: Facts and Politics, 125 pages, CSE, New Delhi.

Dash, M.C. 2010. Climate Change: Facts and Politics (In Oriya language), Bigyan Diganta, 6th issue, 3-14, Orissa Science Academy, Bhubaneswar,

Dash, M.C. 2007. Save the Planet from Global Warming, Society of Geoscientists & Allied Technologist Bulletin, 8 (2), 43-59.

Fatesingh, Himansusekhar, 2010. Energy Generation and Conservation (In Oriya), The Samaj, 26th December Issue.

Forest Survey of India Report, 2009, Government of India, MoEF, New Delhi.

Jain, M.K.2010. Suitability of Fly ash as a mine void- fills material-A critical Review, Minenvis, 64, 1-6.

Kishwan, Jagdish, Pandey, Ran Dadhwal, V.K.2009. India's Forest and Tree cover: Contribution as a Carbon Sink, Indian Council of Forestry Research and Education, Dehradun, 1-12.

Melkania, N.P. 2009. Carbon Sequestration in Indian Natural and planted forests, Indian Forester, 380-392.

India: Taking on Climate Change, Move, Government of India, New Delhi, 6 February, 2010

Paribesh Samachar, 2010. Biodiversity-Ecosystems Management and the Green Economy, Volume XV and XVI, 1-25, State Pollution Control Board, Orissa, Bhubaneswar.

Rain Forests: The Burning Issue.1996 .The Prince's Rainforests Project, Forest Stewardship Council, London, 48 pages.

Sharma, S., S.Bhattacharya and a, Garg.2006.Greenhouse gas emissions from India: a perspective, Current Science, 90 (3), 326-333.

Shukla, P.R.2006. India's GHG emission scenarios: Aligning development and stabilization paths. Current Science, 90 (3), 384-395.

Singh, Gurdeep.2009. Environmental Assessment of Fly Ash from some Thermal Power Stations for Reclamation of Mined Out Areas, In 'Proceedings of Indo-Italian Conference on Emerging Trends in Waste Management Technologyies', Organized by MIT College of Engineering and Maharashtra Institute of Technology, Pune, pp 38-45.

Part IV : Global Warming and Climate Change

12

Is there an Upper and Lower Limit for Carbon dioxide in the Atmosphere for Biodiversity to Function in the Context of Climate Change?

Introduction

Is industrialization the key factor for warm climate in India in general & Odisha in particular? Is it due to the loss of forest cover in many districts? Is it related to changes in the oceanic environment in Bay of Bengal, and Indian Ocean? Is it due to the increasing transport facility? Is it due to the interaction of all these factors? Is it the natural process which scientists have not been able to decipher adequately? Will scientific analysis support the notion that industrialization & increased transportation is the sole factor for climate change? Is the available GHG sink factor in the country not adequate to absorb the GHG emitted from all sources?

This paper makes an analysis with regard to the limits of carbon dioxide (green house gas) in the atmosphere for the biodiversity and human beings for a comfortable living and its relation to global warming.

Green House Gases (GHG) :

Carbon dioxide is a colourless, odourless, nontoxic gas and essential for green plants for photosynthesis and along with water vapour and clouds, this gas keeps the atmosphere warm by absorbing radiation making living processes comfortable for most of the organisms. However it is a green house gas and in recent times has drawn huge attention.

I am a strong supporter of nature conservation, and to live in unpolluted clean environment. The traffic situation, especially road conditions, infrastructure available in road crossings, discipline on the road, civic sense of people, etc is very poor in India and Odisha in particular, even at Bhubaneswar - the capital city of Odisha. There is no foot path, road crossing zones and markings, cycling paths etc in Bhubaneswar and other urban centres of India. Traffic congestion is heavy. Implementation of rules and laws etc is inadequate.

Pollution level due to suspended particles, respiratory particles, carbon monoxide level in the road crossings, pathogens in the ambient air must be in the alarming state. These aspects appear to be equally important with green house gas level in the atmosphere in the country.

The living body is made up of carbon and it is the living stuff in all organisms. Daily, a human being exhales on the average around 1 kg of carbon dioxide and thus in our state, we exhale on the average not less than 41, 000tons of carbon dioxide per day (~ 15 million tons of CO_2 per year) taking 41 million people live in Odisha (as per the 2011 census). Considering the Indian population of ~ 1210 million as per 2011 census, 1210 million kg (1.21 million tons per day, ~ 442 million tons per year) of CO_2 goes to the atmosphere daily by human exhalation.

Human tolerance to CO_2 level

Human beings and most of the animals will survive and will perform their functions without having much CO_2 in the air but with higher concentration of CO_2 in the atmosphere, physiological functioning, especially respiratory gas exchange would be very difficult (Happer, 2011). In otherwards, there may not be a lower limit but there is an upper limit that man and most of the vertebrate animals can tolerate.

NASA, USA have conducted extensive studies on human tolerance for CO_2 for astronauts and NASA recommend an upper limit of 5000 ppm of CO_2 for space missions for one thousand days, assuming total pressure of one atmosphere. USA Navy recommends an upper limit of 8000 ppm for cruises of ninety days in the submarines. (Happer, 2011).

Photosynthesis & CO_2 level

Scientific studies show that the level of CO_2 in the atmosphere should not be less than 150 ppm for plants to continue the process of photosynthesis. However, plants usually grow better and produce better flowers and fruits at higher level of CO_2 (Happer, 2011). The present level of CO_2 in the atmosphere is conducive for good photosynthetic activity.

Evolutionary history says that most of the green plants evolved at CO_2 levels of several thousand ppms, not less than 10 times higher than the current level in the atmosphere. About 57 million years before in the Eocene period of Cenozoic era, the volcanic activity generated huge amount of oxides of carbon and other gases as the geologists and evolutionary biologists tell us (Kump, 2011). All forms of animals and many flowering plants and ancestors of modern man had already evolved in the planet by that time as scientists tell us (Table 1).

Was the high concentration of CO_2 detrimental to living organisms at that time? It may be that the organisms were adapted to high temperature and the biodiversity was not as rich as it is now. But the fact remains that the adaptive potential of biodiversity, especially human beings is phenomenal.

Green plants are, however will do their functions more efficiently in increasing CO_2 level in the air up to a very high concentration with other environmental conditions remaining as it is at present. For us, the important issue is the productivity of crops and productivity of other plants providing food to us under increased CO_2 concentrations. As it appears higher the concentration of CO_2 up to certain level will increase the productivity if the temperature regime, availability of nutrients and water do not create stress. But increasing CO_2 level in the atmosphere will increase the level of atmospheric temperature.

Table 1. The ages of life on earth (based on Reader's Digest Publications)

The Ages of Life on Earth

Age	Events	Life
Precambrian 4.6 billion years ago 4.5 billion years ago 3.8 billion years ago 3.3 billion years ago 2.1 billion years ago 720 million years ago	The Earth is formed The Moon is formed	No Life No Life Single celled organism Cyano bacteria or Blue Green Algae Single celled organism with a Nucleus. Multi cellular Animals.
Paleozoic Cambrian – 550 – 505 million years ago. Ordovician – 505 – 438 million years ago Silurian – 438 – 408 million years ago Devonian – 408 – 360 million years ago Carboniferous – 360 – 286 million years ago Permian – 286 – 245 million years ago		Trilobites & other Marine animals with hard cell. Fishes without jaws & corals. Land plants & Jawed fishes. Land Animals, Insects & Amphibians. Reptiles. Reptiles diversify.
Mesozoic Triaseric (245-208 million)-Juraseric (208-144 million)- Cretaceous (144-65 million)		Reptiles Evol. Peak. Evolution of Birds & Mammals.
Cenozoic Tertiary period (65 – 1.8 million years ago) Paleocene (65 – 57 million years ago) Eocene (57 – 34 million years ago) Oligocene (34 – 23 million years ago) Miocene (25 – 5 million years ago) Pliocene (5 – 1.8 million years ago) Pleistocene (1.8 million years ago) to 50000 yrs, Holocene Quaternary Period (15,000 years ago to present)		Birds diversify. Flowering plants, Mammals & Primates appear Ancestor of Modern man. Other Mammals. Homo erectus Human evolution ,Homo species spread Homoneanderthalensis,H.cromagnon,H.sapiens Modern man.(H.sapiens sapiens)

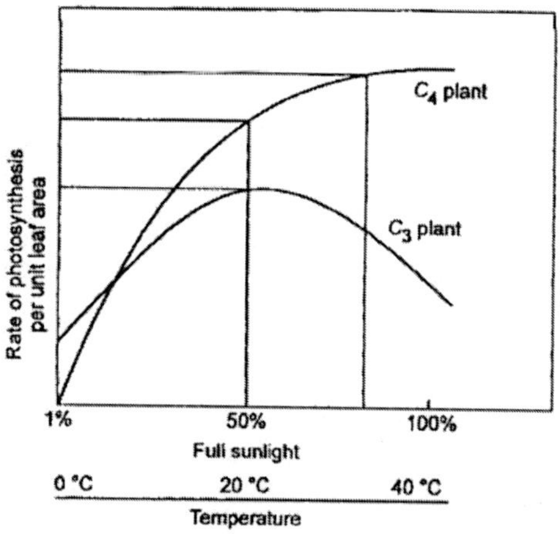

Relationship between intensity of sunlight, temperature and rate of photosynthesis in C_3 and C_4 plants. Rate of photosynthesis for unit leaf area is significantly higher in C_4 plants in different sunlight and temperature conditions

Figure 1 shows a generalized graph showing the relationship of the intensity of sunlight, air temperature and rate of photosynthesis in C3 and C4 plants (Dash and Dash, 2009).In high air temperature, the rate of photosynthesis goes down, especially in C3 plants (most of the plants and crops).

IPCC shows serious concern for rise of temperature in the atmosphere and in the ocean. Besides the rise of temperature is thought to be the cause of glacier melting, loss and migration of biodiversity northward, loss of productivity, flooding of coastal ecosystems and human settlements.

Increasing CO_2 level in the Air

Before the industrialization in Europe, the average CO_2 concentration in the atmosphere was around 280 ppm and during the last 200 years, the concentration has increased to about 395 ppm (115 ppm increase). The current rate of burning of fossil fuels, the rate of increase in CO_2 concentration has been estimated to about 2 ppm per year. At this rate of increase, it would take 300 years for CO_2 to reach a concentration of ~1000 ppm in the atmosphere. The IPCC report (2007) shows that during this 150-200 years post industrial period, the average temperature of the atmosphere has increased to about 0.8 degree Celsius. They also predict that with the current rate of increase in CO_2 and other green house gas emissions in the atmosphere, the temperature will increase by 2-4 degree Celsius, or more, which will be detrimental to many aspects of our crop productivity, prevalence of disease, housing, displacement of people from coastal areas, security, etc especially in tropical countries. Air temperature rise may enhance crop productivity in temperate climate but glacier melting and appearance of vector borne disease will increase.In view of this they advocate to change to alternate energy sources urgently so that CO_2 and other green house gas emissions is reduced and impact would be moderated.

Many scientists are of the opinion that since the Paleozoic era (~ 550 million years before), it is assumed that the ocean currents involving the fluid flow pattern and salt belt, the continental drift, massive volcanic eruptions in initial periods, the planet's orbital parameters (Milanokovitch cycles), variations in sun's output of all types of radiations, and some other causes were the reasons for temperature increase making the planet conducive for life. Besides the atmospheric gases played important role in the heat balance of the planet.

In recent times scientists have found a strong correlation between CO_2 concentrations and planet's temperature by studying ice-core records of the cycles of glacial and interglacial periods during the last million years or so. The planet's climate is always changing and there may be many reasons as discussed above. About twelve thousand years before there was warming of the planet called medieval warming followed by the "little ice age" which brought famine as the crops were destroyed due to severe cold

(Kump, 2011). Since the little ice age, the planet is warming up slowly and steadily. But during the last 200 years, the level of green house gases, particularly CO_2 level has increased and increasing due to human activities. Therefore every nation is willing to adopt 'green technologies' in spite of the fact thc technologies are available in few countries. In view of this green technology development and transfer is now considered as thrust areas of research and climate change activity.

International Scenario

Many private enterprises, especially in industrially developed countries have invested heavily on green technologies and their role in promoting climate change programmes can not be ruled out. However most of the countries, particularly the developing countries such as India, China, Brazil, Mexico.South Africa, Argentina etc depend upon coal for their energy.

The leaders of green technologies are the industrialized western countries, (west European countries, USA, Canada) Australia and Japan and hence the world environmental politics involving development and transfer of green technologies depend upon development of acceptable international protocols.

Considering the scientific facts as discussed above, it can safely be concluded that CO_2 level in the atmosphere is one of the factors of global warming and not the sole factor. The serious environmental issues are the habitat loss and damage to regional/local landscapes, increasing pollution load, particularly hazardous chemicals and wastes, mining activities and unplanned industrialization and massive transport system. Although carbon dioxide level of 395ppm or higher level does not create hindrance in the functioning of the living world, but it absorbs and retains heat and increase the temperature of the atmosphere. The temperature rise creates problem for the living and nonliving world and biological and physical processes.

Energy Scenario

India's Energy Source

About 70% of our energy source (1, 70, 000 MW) is fossil fuels and ~50 % is coal based.

Coal based thermal power plants are not environment friendly as these plants generate huge amount of GHG and pollutants such as SO2, NOx, Cox,, SPM, Fly & bottom ash, metallic dust, etc

. About 120-150 million tons of fly ash is generated annually in India (Jain, 2010). The utilization is only 25%-30% in brick making, road construction, forest soil fertilization, and in agriculture etc (Jain, 2010). Hence it is a huge problem.

Industrial GHG Emissions:

Industrialization has grown manifold in India after 1947.An estimate of GHG emissions and the basis of estimations are given in Table 2.

India annually burns > 300 million tonnes (Odisha > 30 million tons) of coal in power plants and generates equal amounts of CO_2 from this source.

Table 2.The basis of estimations of GHG emissions from energy sector and important industrial sectors in the country with the capacity of production

Energy sector: Thermal Power:

1MWH generation requires about 17 tons of Indian Coal, which contains 32% C. 1MWH power generates about 0.79t CO_2 .Hence 1000 MW, generates about 7 million tons CO_2 per year. India generates ~ 119, 000 MW thermal powers annually and emits about 825 million tonnes of carbon dioxide.

Industrial Process sector:

Iron & steel production: 1 tonne production of steel generates about 2.6-3.5 tons of CO_2 (for sponge iron- about 2.6 ton CO_2) from coal & coke. India produced ~ 44 million tonnes of iron & steel in 2006 with about 10% annual growth. The estimate of current production is about 65-70 million tonnes of iron and steel per year (CO_2 generation is ~210 million tonnes at present). Future projection is > 100 million tonnes of iron and steel by 2020.

Ferro-Chrome:

1 tonne production generates about 2.1 tonne of CO_2. The current Ferro-Chrome demand is estimated as 8 million tonnes (8 x 2.1 = 16.8 million tones of

CO_2 generation). (Taking the values of 84%C and 50% C in metallurgical useable grade coke and coal respectively).

Alumina & Aluminum:

1 tonne alumina production using about 4 tons bauxite ore and 1 tonne aluminum production using about 2 tons of alumina generate about 22 ton CO_2. India has the capacity of ~ 4 million tonne aluminum production per year. (CO_2 emission amounts to 88 million tonnes)

Cement:

1 ton production generates about 0.9 ton CO_2. In 2003, India produced about 117 million tonnes of cement. At present, India has ~ 200 million tonne annual production capacity. (180 million tonnes of CO_2 emission)

There is increased need of energy in the farm, industrial and service sectors as the population continues to grow and to meet the developmental needs. Modern way and standard of living demand huge energy from unsustainable consumerism, transport facilities and increased construction work and other aspects of human use.

Energy is the key for survival, continuance and for sustainable development.

It has been estimated that by 2020, the energy production capacity of the country may reach 220, 000MW, although per capita energy consumption in India is <500 units per year compared to about 2600 units at international level in industrially developed countries.

THE GHG SINK POTENTIAL

Forest and tree cover of the country was 78.37 million ha in 2007 (FSI, 2009), which was 23.84% of the geographical area and this includes 2.82% tree cover.

The National goal is to achieve 33% of forest cover in India. The situation in different regions is not very encouraging. The forest cover is to be increased and special efforts are to be made. Awareness should be created in a massive manner to reduce the carbon foot print of individuals, and industries. We have to give importance on priority basis to renewable sources so that the pollution load and green house gas load to our atmosphere is lessened.

Situation in Odisha- A Case Study

Considering the abundant mineral deposits and water availability, and increasing transport facility, it is natural that many industrial houses

Table 3a & b gives the forest and tree cover in the country in 2007 (FSI, 2009)

Table 3a Forest & Tree cover in India (2007)

Class	Area (million ha)	% of Geographical Area (G.A.)	
Forest Cover			
VDF (very dense)	8.35	2.54	Note: 55.51 million ha lies
			<1000m, 11.67m.ha.
MDF (medium dense)	31.90	9.71	>1000 to 3000m and
			1.91 m.ha >3000m
OF (open)	28.84	8.77	altitude.
Total Forest Cover	69.09	21.02	
Tree Cover	09.28	02.82	
(Tree patches<1ha			
With canopy density >10%)			
Total Forest & Tree Cover 78.37		23.84	
Non-forest: Scrub	04.15	01.26	
Non-forest	255.49	77.72	
Total G.A.	328.73	100	

Forest and tree cover of the country was 78.37 million ha in 2007 (FSI, 2009), which was 23.84% of the geographical area and this includes 2.82% tree cover.

Table 3b.Forest & Tree cover (sq.km) region-wise in India9based on FSI, 2009)

Region	Geog.Area(GA)	Total Forest Cover	% of GA	Scrub
(No.of States/UT)				
South (5)	6, 36, 263	1, 18,284	18.59	14,891
North (9)	7, 72, 644	83,126	10.76	4,561
West (4)	8, 49,676	80,205	09.44	10,272
N-East (8)	2, 62,179	1, 73,182	66.05	919
South East (3)	3, 24,153	82,060	25.32	5,544
Central (2)	4, 43,436	1, 31,878	29.74	2,263
Islands (4)	8,884	6,8 83	77.48	03
Total	**3,287,263**	**6, 77,033**	**20.60**	**38,475**

are interested to invest in Odisha. Coal is our main energy source. It is abundantly available in Odisha and therefore we will depend upon coal for many decades. Coal in Odisha on the average contains 32% carbon and 40- 45% ash. For thermal power generation, about 17 tons of coal is required per day for production of 1 MW. The pollution load from Odisha- coal based thermal power generation amounts to (Paribesh Samachar, 2010) :

~19 tons of CO_2 per day .MW

>136 kg of SO2 per day.MW

>7 tons of fly ash per day.MW.

Currently > 7600 MW of thermal power is generated from different sources in Odisha (Dash, 2010). In view of this the pollution load is heavy amounting to 363861 tons of SO2 emissions per year, ~19 million tons of fly ash per year, and 53 million tons of CO_2 per year. Besides, huge quantity of NOx gases, aerosols etc are pumped to the atmosphere in Odisha.

Out of 30 districts, in the state, only 12 districts have forests cover exceeding the National target of 33% forest cover for each district. Another 4 districts have forest cover exceeding 23% and less than 33% and 13 districts have forest cover less than 20%.

Some districts such as Bhadrak, Jagatsinghpur, and Puri have forest cover of less than 5% and districts such as Baleswar, Kendrapara and Jajpur have forest cover less than 10% of the geographical area of the district Angul and Denkanal districts and many are in the pipeline. These are important districts from population density and industrial development point of view.

The districts of Angul, Denkanal, Jharsuguda, Sundergarh and Sambalpur are under severe stress from industrial emissions. More than 61% of thermal power of the state is generated by industries located in of industries and expansion and conservation of forest cover should get top priority in the state government development policy and should be reviewed periodically.

It is right time to debate whether all green house gases should be declared as pollutant or should be included in the generic structure of Environmental Assessment document (EIA) in India? The carbon foot print of new industries keeping international and national standards in view may form part of the EIA document.

Societal goal for energy consumption should change from high carbon systems to low carbon systems. Observation of environmental discipline at individual, family, societal and national level is the call of the time. In view of this good governance, judicious implementation of environmental laws, and implementation of massive awareness programmes are required.

References

Dash, M.C. 2010. Environment, Energy & Development from Stockholm to Copenhagen and beyond the Celebrations. The BioScan, Special issue, Vol. 1:1-11.

Dash, M.C. 2010. Carbon Sink Potential of Forests and Industrial Green House Gas Emissions, Society of Geoscientists and Allied Technologist's Bulletin, 11 (2), 1-13.

Dash, M.C. and Dash, S.P. 2009.Fundamentals of Ecology, Tata-McGraw-Hill Publications, New Delhi, 562 pages.

FSI Report, 2009.Government of India, New Delhi

Happer, William.2011. The Truth about Green House Gases, First Things, (USA), 33-38.

IPCC, 2007. A Report of the Intergovernmental Panel of Climate Change, 2007. (4 volumes), UNEP, Switzerland

Paribesh Samachar, 2010. SPCB, Odisha Publications, Bhubaneswar

13

Impact of Climate Change on Natural and Man-made Ecosystems

Introduction

Before understanding the impact of climate change on natural and man-made ecosystems, it will be convenient to understand the meaning of a system and what an ecosystem is? And what services they provide with to the mankind.

What is a System?

A *system* is a set of interacting or interdependent component parts forming a complex or intricate whole. Every *system* is delineated by its spatial and temporal boundaries, surrounded and influenced by its environment, described by its structure and purpose and expressed in its functioning.

What is a Living System?

Living systems are open self-organizing *living* things that interact with their environment and interact with individuals of same organism and between other organisms. These *systems* are maintained by flows of energy and matter. The microorganisms, plants and animals of a habitat & their nonliving environment function together as these are connected to each other through food chain (eating and being eaten), matter (sodium, potassium, calcium, phosphorous, sulphur etc) circulation, energy flow etc. as one unit, which is called as an Ecological system or Ecosystem, coined by A.G.Tansley, British Ecologist in 1935. Ecosystem concept is holistic because all component parts of the ecosystem function together (if one component part shows malfunction then the entire system fails) and therefore, has functionally obligatory relationship & Man is part of the ecosystem. A natural pond, Lake, Bay, Ocean, Grassland, Forest, etc. are individually considered as natural ecosystems.

The component parts of an ecosystem: (i) Producers' (green plants), which transduct solar energy and produce carbohydrates in a biochemical process called photosynthesis, and they produce food for other biodiversity and for man; (ii) Consumers such as herbivores, who graze over plants, living or dead, and Carnivores, (called C1) who predate on herbivores and large carnivores (called C2) predate on herbivores and on small carnivores, and (iii) Decomposers decompose organic matter and in the process materials are released for use of the producers (plants) to grow and produce carbohydrates in photosynthetic process; (iv) the abiotic (non-living) matter such as some minerals and organic nutrients and factors such as temperature, and other environmental factors.

The structure of an ecosystem refers to the above cited components and to the species diversity, their distribution, numbers, and biomass in the ecosystem, the qualities and distribution of organic and inorganic nutrients and the physicochemical conditions characterizing the system.

Ecosystem function/ Ecosystem services refers to the energy flow and its rates, nutrient cycling between the biotic and non-biotic components, biogeochemical cycles such as water cycle, oxygen and carbon dioxide and other gaseous cycles, and mineral cycles, such as calcium, potassium, sodium, sulphur etc and the homeostatic mechanisms that determine the structure and stability of the ecosystem.

The natural ecosystems of the world include Tundra, Desert, Lakes, Rivers, Ponds, Oceans, Forests (Tropical, Subtropical, Mangroves, Temperate, Alpine, etc), and Grasslands such as prairies of Canada, in different latitudes and Altitudes and they provide with all necessities (food, fodder, shelter materials, ayurved medicines, raw material for pharmaceutical processes to prepare allopathic medicines, oxygen for all aerobic organisms including man etc) of human beings and to other species.

These ecosystems may vary with respect to species composition at producer, consumer and decomposer levels but they perform similar functions.

What are the component parts of a forest/grassland ecosystem? What are the component parts of the aquatic ecosystems? Are the producers, consumers such as herbivores and carnivores, and decomposers are same in these ecosystems? What is about the functional aspects? What are the ecosystem services provided by these natural ecosystems? How the natural ecosystems can be utilized to increase food, fodder and fuel, production and resource for ayurved and pharmaceutical science and to protect environment (landscape) from erosion, pollution etc. and manage the resources? This is a big question mankind faces in the context of climate change.

India is one of the eighteen mega biodiversity countries of the world having about 8% of world's biodiversity although land area is only 2.4% of the planet.

The Himalayan ecosystems & the Western Ghats ecosystems arc international biodiversity hot spots .There are 26 Ramsar Sites (aquatic sites) in India. They harbour rich biodiversity and provide invaluable ecosystem services to mankind.. The Chilika Lake in Odisha was the first Ramsar site declared in Ramsar Convention, held in Iran in 1976 for the protection of aquatic ecosystems and biodiversity, especially water fowl.

The functioning of natural systems is based on certain basic principles and understanding of these principles will help man to solve many of its own problems.

Ecosystem Services

All forms of life owe their existence to the mother Earth and to the maintenance of ecological order (ecological balance) in nature.

This balance is maintained by: (i) Rich bio-diversity, (Microbes, Plants, Animals, & Man), (ii) Interdependence among the different life forms, (iii) Sustained productivity and food chain relationships, (iv) Maintenance of natural processes of nutrient cycling, energy, flow, homeostatic mechanisms and evolutionary process, (v) Adapting potentiality of the living forms to the changing environment.

A forest ecosystem combats climate change. They absorb excess and harmful carbon dioxide (CO_2) from our atmosphere. In fact, in a single year, an acre of mature trees absorbs the same amount of CO_2 produced by driving the average car 26, 000 miles. Forests help us breathe clean air. Forests absorb odors and pollutant gases (nitrogen oxides, ammonia, sulfur dioxide and ozone) and filter particulate matter out of the air by trapping them on their leaves and bark. All the living forms including man get their requirements of gases such as oxygen, other matter, and energy from the ecosystems where they live.

Forerst ecosystems help mankind to achieve long-term economic and environmental sustainability and provide food, energy and income. They enrich soil with nutrients and carbon. We can scan different species of trees for higher C-sequestration potential & recommend plantations. We can study carbon -sequestration by forest soil fauna and flora and suggest their maintenance & conservation.

The biosphere consisting part of hydrosphere (water), pedosphere (soil), atmosphere (air) and the living forms (bio-diversity) have arisen in the evolutionary process. The living microorganisms, plants and animals have specific requirements for existence, continuance and perpetuation.

Ecology, a word derived from the Greek word Oikos, meaning house, is concerned with the biology of organisms, populations, communities etc. and their functions in natural surroundings. In another words, ecology deals with the understanding of the structure and function of nature; this includes the living world including man and the nonliving substances, which are constantly taken by the living organisms from their surroundings to keep them alive.

The practical importance of ecosystem analysis results from the presence of man in the ecosystem. Man's evolution, its dependence on environment, his relationship with nature and the development of his position from one of struggle against natural forces to the recent influence and use and abuse of it are included in ecology and ecosystem analysis.

Pre-agricultural human society, agriculture- based society, industrialization, urbanization, the atomic age and the electronic age, and the global warming and climate change have brought the science of ecology to the forefront of human knowledge.

The green plants are the autotrophs. The consumers & decomposers are the heterotrophs. The green plants, the consumers such as herbivores and carnivores, the detritus feeders such as bacteria, fungi, and soil invertebrate animals form the biodiversity of an ecosystem. Man is also part of the ecosystem. Photographs of natural ecosystems and list of biodiversity found in ecosystems are given below.

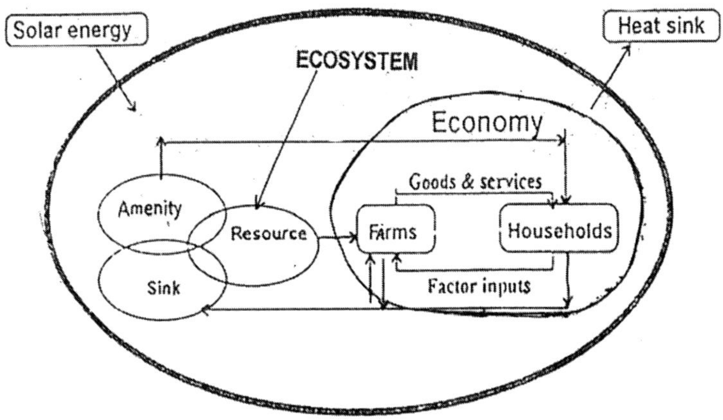

Ecosystems

Man-made Ecosystems

Agriculture is a man-made ecosystem, so also man-made fishery ponds. Agriculture involves usually monospecies crops such as paddy, wheat, maize, barley etc. Many cultures required species of fish for quick growth for his consumption. What are the component parts of these systems?

Differences between natural and man made ecosystems

The natural ecosystems are largely pollution free and less energy intensive systems as they depend upon solar energy. These are multispecies, rich biodiversity, stable, with very complex food webs systems.

Manmade ecosystems are the Agro ecosystems, Aquaculture systems, Industrial systems, Urban systems etc. These systems may be fragile and require huge amount of energy; usually fossil fuels, resources, human labor or machines for their functioning. These systems, however provide jobs and productive, although they generate pollutants that create health hazards.

Impact of Climate Change on Ecosystems

Climate Change occurs due to global warming (rise in earth's temperature). Global warming is linked to the concentration of green house gases in the atmosphere and their global warming potential and residence time. The concentration of green house gases in the atmosphere depends upon the emissions of green house gases by human activities, especially by burning fossil fuels and natural processes. The impact of climate change has already been seen with occurrence of extreme weather events; very heavy rain falls and rains falling in few hours causing heavy floods or extreme droughts extending for considerable time. The impact is marked on agriculture, biodiversity, heat waves, human health, sea level rise, glacier melting and many other ways that affect our living process. Some of these impacts are narrated here.

Impact on Agriculture

Most of the south-east Asian countries including our country have an agrarian based economy. Large per cent of our population are engaged in agriculture including livestock rearing and management. Climate change affects agriculture in a number of ways, including through changes in temperatures rise and heat waves, heavy or scanty rainfall, and changes in the occurrence of pests and diseases; changes in atmospheric green house gases, especially carbon dioxide and ground-level ozone concentrations; and changes in sea level and sea water ingression to land.

Climate change is already affecting agriculture in many countries, with effects unevenly distributed across the world. Future climate change will likely negatively affect crop production in countries, situated in low latitude. The effects in northern latitudes may be positive or negative. Climate change will probably increase the risk of food insecurity for poor people and weaker section of the society.

The Intergovernmental Panel on Climate Change (IPCC) has published several reports and the report published in 2001, concluded that the poorest countries would experience with reductions in crop yields in most tropical and sub-tropical regions due to decreased water availability, and new or changed insect pest incidence. In Africa and Latin America the yields of many rainfed crops are likely to fall sharply for even with small climate changes as these crops may not tolerate high temperature;; and hence reduction in agricultural productivity of up to 30% over the 21st century are projected. Marine life and the fishing industry will also be severely affected in some places. A 2008 study published in *Science* suggested that, due to climate change, "southern Africa could lose more than 30% of its main crop, maize, by 2030. In South Asia losses of many regional staples, such as rice, millet and maize could be higher than 10%".The average crop yield is expected to drop down to 50% in Pakistan whereas corn production in Europe is expected to grow up to 25% in optimum hydrologic conditions.

More favourable effects on yield tend to depend to a large extent on realization of the potentially beneficial effects of carbon dioxide on crop growth and increase of efficiency in water use. Decrease in potential yields is likely to be caused by shortening of the growing period, decrease in water availability and poor vernalization and high atmospheric temperature.

Impact on Biodiversity

Plants and animals which form biodiversity are affected by floods, droughts and extreme weather conditions. Tigers (*Panthera tigris*) have lost their natural habitats in Sunderbans due to sea level rise, and in many other forests due to loss in forest cover in our country and have become endangered. The Impacts may include: (i) Extinction; (ii) changes in species ranges; (iii) Phenology; timing in flowering, pollination etc (iv), (iv) Decline in populations/ some species may flourish.

There is uncertainty on how biodiversity will adapt quickly to the climate change associated with other variables- it may be that migration will help to adapt, but with habitat loss/ ecosystem degradation, & wildlife-Human conflict, migration appears to be very difficult. Climate change is happening rapidly and hence adaptation time is minimal. Food web relationships breaking due to temperature rise & other factors and bringing huge changes in biodiversity.

Global warming has visibly affected the winter hibernation period in many animal species, seriously affecting their metabolism, food requirement and life history. Some of the examples are listed here:

ROCKY Mountains, USA-Yellow bellied marmots are emerging 38 days earlier from hibernation. Ground squirrels emerging 10 days earlier than the average data of 20 years exhibit. Hedgehogs in U.K.are emerging 21 days earlier from hibernation. In Florida, hibernation of Bats has been affected; they emerge earlier as spring temperature increase. Hibernation period has been reduced in pregnant females, who usually store energy before winter and energy storage has been reduced in Canada & Alaska. Brown Bears in Cantabrian Mountains in Spain do not hibernate. The Himalayan Brown Bear & the Asiatic Black Bear are not hibernating in winter. Climate change is affecting hibernation cycles of many frog species including the endangered Baw-Baw frog in Australia. Spotted Snout Burrower Frogs breaking their hibernation period depending upon seasonal rainfall in South Africa. (Source: www.downto earth.org.in).

Due to global warming and climate change few wild edible plants and edible mushrooms run the risk of extinction in our country. Out of many varieties of mushrooms found in Odisha, Amba Chhatu & Bali Chhatu has become rare because of erratic rainfall. The tribal people living near the Andhari Landscape-sacred grove in Jharsuguda district used to sell these mushrooms for their livelihood (plate).

Some other examples are: Makall beru (*Dacalepsis hamiltoni* L), a climber found in scrub and dry deciduous forests of Karnataka. The Soliga tribes are used to eat the roots of this plant and also sell the roots to earn their livelihood. Six years ago, harvesting of roots has been banned to save this species. Another example is the *Khyndong hati, a conical mushroom found in Meghalaya forests on logs of dead wood is a popular food among the people of jaintia hills is disappearing due to forest loss and climate change. Maitenga (Passiflora assamica) is a wild edible plant in Assam and the leaves of this plant is cooked with small fish as a delicacy. This plant species is disappearing due to forest loss and climate change.*

Out of many varieties of mushrooms found in Odisha, Amba Chhatu, & Bali Chhatu have become rare because of erratic rainfall

Coral reefs provide niches to much variety of fishes and marine organisms. Because of global warming corals are bleaching. The following plate shows the situation.

Great Barrier Coral Reef-21.11.2016- 30.11.2016. Breeding Season for Corals (Coelenterate- Polyps produce eggs and sperms in large numbers-fertilization-plannula larva-. (with right temp., salinity & pH level) -Yahoo Home Page

Impact on Health

Global warming is changing the breeding pattern and ecology of mosquito including disease spreading mosquitoes of Malaria and Dengu and other insect vectors of disease. The situation in turn has caused serious health issues in human beings. The World Health Organization has recently published a chart linking climate change to human health (Figure). Heat waves are causing more deaths and sick people in summer, especially in tropical and subtropical regions.

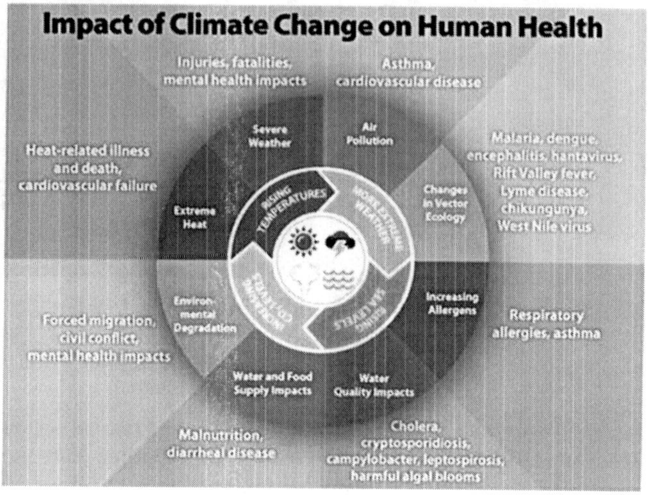

Impact on Glaciers

One of the most dramatic impacts of global warming is seen on the glaciers. Himalayan Glaciers have in recent past receded at an alarming rate (Table and plate). If we lose our glaciers, it will affect the water flow in perennial rivers such as Ganga, Yamuna and Indus. This situation will bring miseries to more than 200 million people, who depend upon these rivers.

ISRO Glaciers Receding

Sl. No.	*Glacier*	*Basin*	*Glacier Area SQKM*	*Ret Reat Total*	*In Metres Rate/Yr.*
1.	Miyar		87.8	757	16.43
2.	Shaune Bage	Satluj	6.8	923	26.4
3.	Bilare Bange	Satluj	2.8	90	2.6
4.	Parbati		48.44	5569	214
5.	Samudra Tapu		77.67	302	21
6.	Chipa	Dhauliganga	5.0	1050	26.92
7.	Meola	-Do-	14.0	1350	34.62
8.	Jhulang		143	535	28.1
10.	Dokariana Bamak	-Do-	5.8	585	16.7
11.	Meru Bamak	-Do-	4.7	395	17.2

Himalayan Glaciers melting, MoEF, New Delhi

Impact on Sea Level Rise

The rise sea level is expected in future. The island nations and coastal cities such as Dhaka, Mumbai, Kolkata, and Tokyo, New York are under threat and the millions of people living in coastal areas.

Further Reading

Dash, M.C. and S.P.Dash.2009. Fundamentals of Ecology, McGraw-Hill Publications, New Delhi, 562 pages.

Down To Earth, 2017. State of India's Environment, Centre for Science & Environment, New Delhi, 226 pages.

Down To Earth, 2015.State of India's Environment, Centre for Science & Environment, New Delhi, 208 pages.

Down To Earth, 2009.Climate Change: Politics and Facts, Centre for Science & Environment, New Delhi, 125 pages. www.downto earth.org.in

Google.com.search: Impact of Climate Change on Agriculture, Wikipedia, 2017.

Horn Bill, BNHS, Bombay, 2015-2017 issues.

IPCC Publications on Climate Change, 2001-2014. UNO, New York/Geneva.

The Climate Reality Project, India.2016. Understanding Climate Change, UNESCO, New Delhi Office, 4 pages.

14

Recent Trend in Climate Change – International and National Scenario from IPCC

Introduction

Forty five years before on 5th June 1972 at Stockholm, the world realized the future of mankind is bleak unless we take urgent actions to protect our environment from the colossal degradation and damage we have inflicted on the planet in the last few hundred years and pledged to bring worldwide awareness among the people to protect the environment, the mother Earth for our survival. The Stockholm conference lead to the establishment of UN Environment Programme (UNEP) and gave emphasis on protection of environment for the future.

The World Commission on Environment and Development was established in 1983 under the chairmanship of Dr. Gro Brundtland, a distinguished physician and the then prime minister of Norway and brought out its report 'Our Common Future' in 1987.This report highlighted on poverty, depletion of natural resources faster than exploitation, deterioration of environmental quality and quality of life. In 1988, the Intergovernmental Panel on Climate Change (IPCC) was formed by UNEP and published its first report on climate change in

1990 and supplement in 1992. These actions lead to the organization of the Earth Summit in Rio de Janeiro in Brazil in June, 1992.

The Rio conference in 1992 gave widespread visibility of the important concept of sustainable development. This concept placed environmental management in a position to address the environmental impacts of developmental projects so that the environment is protected for the present and future generations. The conference in Johannesburg, South Africa in 2002 provided opportunity to review the actions taken and benefits made and to review the different aspects of resource utilization and conservation. Biodiversity conservation received top priority. Besides, the important issue of energy use and climate change and its effect on natural ecosystems, on the developmental projects, on human activities and future of mankind was very much focused.

The theme in the World Environment Day on the 5th. Of June 2010 was "Many Species, One Planet, and One Earth. The theme respectively in 2009 and 2008 was "The Earth needs us, UNite to combat Global Warming and save the Earth" and 'Carbon dioxide: Kick the habit for low carbon economy'. The UNO has declared 2010 as the 'International Year of Biodiversity Conservation'. The UNO has also declared 2011 as the 'International year for Forest Conservation'.This shows the growing importance of protecting the environment, conserving biodiversity, especially forests to combat climate change. Combating climate change involves the phasing out of the use of fossil source of energy and switching over to other forms of energy, especially renewable energy which do not carbonize the world.

These annual sermons, however regarding the state of environment of earth clearly show that our modern way of living is at stake because the present style of progress and mode of development with unsustainable consumerism and energy source from fossil fuels eats away into our basic life supporting system and puts tremendous adverse pressure on environment, and reduces biodiversity, which provides us food, shelter material, raw material for medicines, gases for survival, other ecosystem services like recycling of water and other materials – something like legendary Indian poet Kalidasa, before being blessed by the goddess of learning was cutting the tree branch sitting over it.

Virtually all earth's ecosystems have been significantly transformed through human activities and the changes have been more rapid during last hundred years as the population grew faster and migration to urban centres became a necessity due to lop-sided development and management. It is now realized that the present concept of unlimited growth in a limited resource based environment is flawed and sustainable development is necessary.

With huge population increase, especially in developing countries and new types of ailment looming on the horizon consequent to a synthetic life emptying the natural medicine tool box consequent to extinction of the medicinal plants is disastrous. Who knows what types of metabolic assaults humanity will face in future in Nature's feed-back loop? In view of this the biodiversity conservation is very important.

We have observed very harsh summer throughout India in 2016.Chennai experienced unprecedented heavy rains in November-December, 2015, flooding the city and causing disastrous situation for the people. In 2016 and 2017, the shortage of drinking water in many parts of the country including Odisha is severe and has brought misery to the people.

UNO Initiatives:

Scientists involved in studying and preparation of the IPCC (Intergovernmental Panel on Climate Change, UNO) climate change reports (IPCC publications, 1990, 1992, 1995, 2001, 2007, 2014) have conclusively proved that climate change is real and its impacts are disastrous. The global increases in CO_2 concentrations are primarily due to-

- Fossil Fuel (coal, petroleum products) Use;
- Land-use and Land-use Change;
- Agricultural activities;
- Industrial Development;
- Forestry and forest loss

The following are the main impacts of climate change and some of these are::

- Temperature;
- Rainfall (Precipitation) ;
- Mountain Glaciers;• Sea Level Rise;

- Health;
- Agriculture;
- Coastal Erosion;
- Biodiversity Loss;
- Storm/Storm Events;
- Soil Moisture Availability;
- Sea Surface Temperature

The adverse impacts are now visible in many parts of the world, especially in south-east Asia including India in the form of severe cyclone, flooding in some areas, drought in some areas, reduced crop production etc. Hibernation period in many vertebrate species is shortened affecting their metabolism and life cycles. Coral reefs are bleaching (BNHS, 2016) and seriously affecting fish diversity and life cycles. This situation has been realised world over since 1960s and efforts have been made by UNO (Table 1) to arrive at principles of taking urgent action by the different nations of the world.

Table 1. Road map to cooler world- UNEP initiatives

- 5th June, to 12th June, 1972 Stockholm – Man & His Environment- UNEP formed
- 1979-Montreal – Climate Change Conference
- 1983-1987-Brundland Commission Report- Our Common Future
- 1988- Intergovernmental Panel on Climate Change (IPCC) was formed
- June, 1992 Rio de Janeiro – Earth Summit- Sustainable Development, Agenda-21-Biodiversity, Climate Change etc
- December, 1997 Kyoto Protocol – Climate Change-Carbon Trading
- November, 1998 a plan of action for implementation of Kyoto Protocol was created under the influence of USA, a cumulative target of 5% reduction by 2012 against 1990 baseline was set. China and India escaped emission reduction target.
- July, 2001 George W Bush pulled out USA from Kyoto protocol citing the non-inclusion of India and China from emission reduction target.
- 2002 Johannesburg- Review
- February, 2005 Kyoto Protocol cane into force, and it would end in 2012.
- December, 2007 Bali Road Map- Climate Change
- 2008-Pozanan Conference- Climate Change

- 2009 Bonn, Germany-Climate Change
- December, 2009 Copenhagen- Climate Change-USA became instrumental in drafting the Copenhagen Accord, which was not adopted. The accord did not commit nations to a binding reduction of GHG.
- November, 2010 Cancun, Mexico- Climate Change Conference.USA pledged GHG emission cut of 17% by 2020 against 2005 And 42% cut emission cut by 2030 confirming to the goal to reduce emission by 83% by 2050.This commitment was lowest for any developed nation.
- 2011 Johannesburg- Climate Change & Sustainable Development
- 2012 Durban- Climate Change
- 26, November to 8, December, 2012 Doha- Climate Change conference- Kyoto protocol was extended to 2015.
- 2012 Rio de Janeiro- Climate Change, Sustainable Development, -
- The Future we want- Common but Differential Responsibility.
- December, 2013, at Warsaw Summit, nations agreed to have national
- Level climate action plans called Nationally Determined
- Contributions (NDCs). The responsibility was on countries on self Differentiation.
- March, 2015.USA submits its NDC with the commitment targets 26-28%
- Reduction by 2025 against 2005 levels.
- December, 2015 Paris, France- 195 countries have participated (Nov.30-Dec.12) - The Accord specified limiting global warming to 2°C rise, Preferably 1.5°C.
- September 3, 2016. USA ratified Paris Accord (President of USA was Mr.Barack Obama). October, 2nd 1016-India ratified the Accord.
- 1, May, 2017. In the Federal Budget, USA (under President Trump) -
- Announced it will not fulfill the Financial pledge of $3 billion to the Green Climate Fund.
- 2, June, 2017. USA pulled out from Paris Agreement calling it unfair & Blaming China & India.

  Note: As per Paris Agreement the pull out will only be effective from 20th July, 2020, - the year of Presidential election in USA.

UNO is persuading nations to take urgent actions on mitigation measures and adoption of different methods to address the reduction of effects of climate change. This matter was discussed in Paris Conference during 30 November to 12 December, 2015. This chapter includes the global warming and climate change problem addressed in different international conferences, especially in Paris Deal.

International Land-mark conferences on the problems of climate change:

In 1997, the global community met in Kyoto in Japan to work out mechanism for controlling Green house (GHG) emissions. Policy decisions taken during this conference have been termed as *Kyoto Protocol*. According to the mandate of this Protocol, all countries were grouped into two categories. Those consuming energy above the global average mostly the industrially advanced countries like European countries, USA, Canada, Japan, Australia, and Russia, are called Annex-1 countries (Developed) and the rest are called Non-Annex-1 countries (Developing). According to Kyoto Protocol, Annex-1 countries have to bring down their GHG emissions to less by 5.2 % average from the base level of 1990 keeping in mind their economic development, population growth, etc. If for some reason they are not in a position to bring down the emission to below 1990 level, they are allowed to go and reduce the GHG emission in the developing countries where the efficiency of energy consumption is also lower and there is a need to consume more energy for ensuring their development. Any reduction made in the specific green house gas (GHG) emissions can be considered as being reduced in the respective country providing the monetary benefit.

Any organization or industry in the Non-Annex-1 country reducing the GHG emissions can trade those emissions with the Annex-1 countries. The Kyoto Protocol further stipulated that the time period allowed for GHG emission reduction commences from January 2000 and ends up by 2012. The accounting for Annex-1 countries begins from 2008 and ends up with 2012 (first crediting period). The process is controlled by the respective governments and the UNO.

To reduce the global warming potential (GWP) from the GHG, industries can undertake any of the following CDM projects: Increase in methane gas (biogas) consumption, Increasing the use of biomass in place of fossil fuel for energy generation, Fuel switching from coal to biomass (biomass gasification), Switching over to hydroelectric power, Energy efficiency projects, Fuel switching from coal to gas, Improving energy efficiency at the user end, Improved Agriculture practices and to increase carbon sink potential.

This financial instrument of selling GHG emission is called carbon trade under Clean Development Mechanism. Kyoto protocol does not necessarily reduce industrial GHG emissions in the world taken in whole, although it aids to aforestation and forest conservation.

Carbon Trading essentially means exchanging points earned by the reduction of GHG for money. This has emerged as a huge opportunity for Developing countries where GHG emissions are much below the quota fixed. Developing Economics are not expected to meet emission targets during this period. Companies in Developing Countries can earn carbon credits by investing in Clean Technology. The principle is based on the fact that climate change is a global problem that requires a global solution. Global cooperation on technology transfer is important and developing countries need access to climate-friendly technology. Global Environmental Facility (GEF) has a critical role to play in the co-development and transfer of technology and it can be made through several different channels. Technology transfer must be accompanied by capacity building. Though UNFCCC (United Nations Convention on Climate Change) has made substantial progress in evolving the mechanism for inter-state sale of carbon emission credits; there are still some gray areas, which are being resolved from time to time.

Carbon trading was in full swing in many developing countries like Brazil, Mexico, China and India despite the fact that USA and Australia did not ratify the Kyoto Protocol.

Carbon trading is moving across all the countries in a very rapid way and expected to involve 1000 billion dollars trading by 2012. The organizations can reduce energy consumption or release of any of the six GHG gases by adopting cleaner options (Clean Development Mechanism) and claim reduction in GHG emission. Renewable and methane captures are at the top of the list and energy efficiency at lower end. As the accounting period approaches (2012) the prices may increase further. Size of the project is determined based on the total emissions saleable over accrediting period.

Developed nations, especially USA, Japan and Australia wanted that large economies like China, India, Brazil, and Mexico should take some steps to control carbon emission although they are listed under Annex-2 countries. This was the scenario before 2007.

Bali Road Map (2007)

UNO organized the conference in 2007 at Bali. The slogan was 'NO TIME LEFT- ACT NOW'. There were 10000 Participants, 187 Govt. Representatives and the conference was held from December-3-14, 2007 and the conference adopted Bali Road Map: a new negotiating process was to be concluded in 2009 & will lead post 2012 and lead to international agreements. USA agreed for Bali Road Map of December, 2007. The salient Features were:

1. Lunching adoption fund (AF),
2. Reducing GHG emissions;
3. Technology Transfer is to made simple
4. USA & Australia agreed for post 2012 agreement & they argued for some commitment from developing countries, especially from China, Brazil and India for reduction of GHG.
5. The Message was 'The World cannot afford to wait'.

Countries are to address the question of a shared vision of long term cooperative action-carbon emission reduction taking into account social and economic conditions and other relevant factors.

Countries must negotiate the question of 'Enhanced action on technology' but countries shrink when clarifying what they wish to do mitigate climate change.

Poznan, Poland-Climate Change Conference (December 1-12, 2008) :

UNFCCC-14th Conference of Parties (COP) to solve the problem of erosion of trust between developed and developing countries on implementation of Bali Road Map was held at Poznan. Over 3000 delegates from 186 countries participated in the conference. Developing countries need huge amount of money every year till 2015 to adapt to climate change effects according to the Human Development Report of UNDP, 2007.At present the availability of the fund is meager and developed countries are not contributing as expected.GEF has been handling money disbursals. Least developed countries (LDC) found the process too much cumbersome. Besides, United Nation's Framework on Climate Change (UNFCCC) has found that Clean Development Mechanism (CDM) process cannot properly account for energy efficiency projects in the building sector.

Policy Shift by USA

There was a policy shift in USA with regard to Climate Change after Mr. Barack Obama became President and Mrs. Hillary Rodham Clinton was appointed as Secretary of State. I quote some of the portions of the address of the Secretary of State Mrs. Clinton delivered at 17 major Economies forum on Energy, Climate on 17 April, 2009.

'It is an environmental issue, a health issue, an economic issue, an energy issue and a security issue. It is a threat that is global in scope, but also local and national in impact.' Desertification and rising sea level generate increased competition for food, water and resources. But we also have seen increasingly the dangers that these transpose to the stability of societies and governments. We see how this can breed conflict, unrest and forced migration. She stresses on four points. These are: 1. Climate change is a clear and present danger to our world and demands immediate attention, the science is conclusive.2. The US is determined to make up for lost time, both at home and abroad and the USA is committed to address this issue and to act., 3. The countries should pull together and work towards a successful outcome of the UN climate negotiations later in the year at Copenhagen, 4. We all must cooperate and new policy and new technologies are needed to resolve this crisis. The USA has made budget provisions for funding and for providing loan for clean energy development, targeting to double supply of renewable energy.

The international business on CDM will be huge; their priority area is to make the homes and buildings more energy efficient so that carbon emissions would be reduced.

On one hand it was very encouraging but there was an element of business interest which makes developing countries worried. They feared that the developed countries may vouch for imposing restrictions on carbon emissions. (Times of India dated 23 May 2009. Every nation wants its enhanced economic development, although shows concern for global warming and emissions.

Bonn, Germany Conference

The first draft for negotiations on the Climate change policy beyond 2012 released on 19th May 2009 at Bonn, Germany included two options for developing economies like India to deviate their emissions from the current base line by 15-30% by 2020 or reduce by

25% from 2000 level by 2050.This generated lot of heat in the negotiations. This provision deviates from UNFCC's original text and also the spirit of Kyoto protocol.

The 20 nation Major Economic Forum (MEF) met in London on 19th & 20th October, 2009 and suggested that International monitoring of GHG should be mandatory and the proposal was not accepted by the developing countries.

Copenhagen: December 7-18, 2009

The Copenhagen conference was attended by 193 countries & 120 Heads of States. The deliberations resulted in developing the following:

- No specific emission targets set for the developed countries
- Consensus was to keep temperature rise to below two degrees Celsius by 2050, but no strong commitment was made.
- Emphasis was on mobilization of financial resources for supporting reforestation efforts of developing countries
- Scaled-up funding for poor countries, including long term cooperative action (fast-tracked fund) of $30 billion during 2010-2012.USA will take initiative to mobilize $100 billion by 2020

Summary of Copenhagen Conference- Proceedings:

- Most of Annex-1 countries wanted a new deal replacing Kyoto Protocol (5.2% cut of 1990 level). Non-Annex countries insisted of retaining Kyoto and action in 2008-2012.Kyoto Protocol survived-discussions to continue
- Consensus (120 out of 193 countries) agreed to allow up to 450 ppm CO_2 & 2oC rise in temp by 2050.
- Outcome was voluntary emission cuts by BASIC countries and retention of Kyoto. (A political agreement was signed- neither supported by UNO not a legal document)
- International monitoring was not accepted.
- Consultations yield a political accord not a legally binding treaty.
- UNO-UNFCCC meeting in Mexico in 2010 (November 29, 2010 to December 9, 2010) did not introduce punitive measures for violation of International agreements. (?).

- Rich nations to take on internationally-binding commitments on reducing GHG (Kyoto, Bali protocols) & Developing countries to take domestic action to cut the GHG emissions- no legal document.
- UNO to play more active role in the governance of flow of finances to help developing countries counter climate change
- Countries agreed to scale up public finances significantly for GHG reduction.

IPCC published their findings (2007) and the expected temperature rise model based on scientific analysis is given in Figure 1 below.

Graph 2.6: Expected temperature rise till 2100

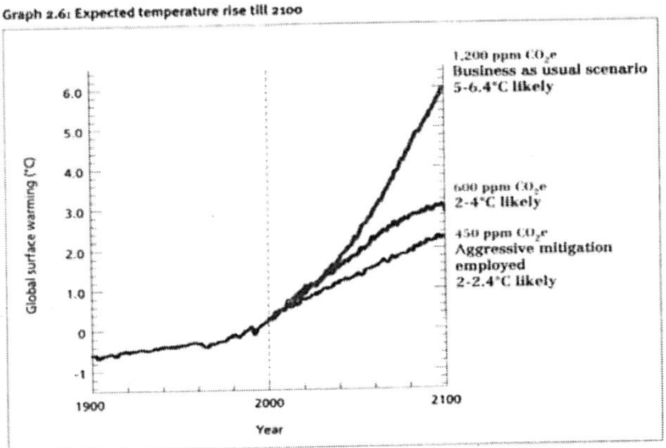

Source: Fourth Assessment Report, 2007, Intergovernmental Panel on Climate Change

Expected Temperature Rise: IPCC

The threshold is estimated as a temperature rise of about 2°C above pre-industrial levels, and on current emissions trajectories, we are heading for a rise of about 4- 5°C without significant mitigation between 2050-2100AD.That may not sound like much, but the temperature difference between today's world & 1900AD is 0.83°C and the difference from the last ice age (about 12 thousand years before) is about 5°C, so seemingly small changes in temperature can mean big differences for the Earth.

GHG Emission Sources

Total GHG Emissions about 36-38 Gigatonnes of CO_2e by 2010 & about 48-54 Giga tonnes in 2014. With 2.3% increase annually. The sources are:

Electricity & Heat Production	25%
Agriculture & Forestry	24%
Industry	21 %
Transport	14%
Buildings	06.4%
Miscellaneous	09.6%

Source: Fifth Assessment Report of IPCC (2014-as Reported by'Down to Earth'-16-30 November 2014, pages 54-55 pages).

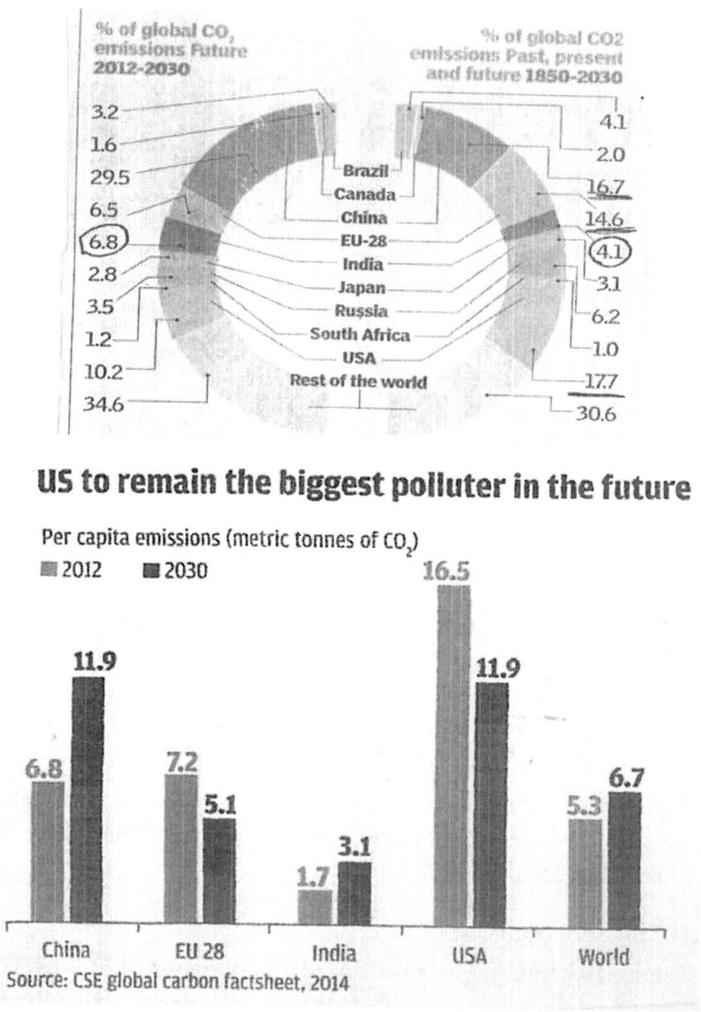

(Figure 2)

The following diagram (Figure 2) shows percentage of global CO_2e emission from 1850 to 2030 and expected emissions from 2012-2030 (Down to Earth 2014). India's per capita emission on the average may go up to 3.1 to 4.1 tonnes per year by 2030 and it is predicted that the emissions during 2012 to 2030 may go upto 6.8 tonnes per year in comparison to 29.5 tonnes in China, 17.7 tonnes in USA and 14.6 tonnes in European Union. (IPCC, 2014 and Government of India Publication, 2015, Down to Earth publications, 2014-2016).

The total emissions per year at present amount to about 36-38Gt and the total emissions will amount to about 748.2Giga tonnes of CO_2e by 2030 AD resulting an accumulation of 2900 Gt in the atmosphere.

India's Voluntary Policy Shift

Considering 7- 8% GDP growth rate per annum in 2002-2010, the estimate of GHG from all sources in 2000 in India was about 1485 million tons, per capita emissions about 1.1 tons. This was about 3% of total global emissions. An assessment of the current and projected trends indicate that green house gas emissions grew @ 4% annum during 1990 & 2000 and projected to grow further due to developmental needs. The absolute level will not be less than 5% of global emissions in 2020. In India, the estimate was 2115 million tons of GHG emission in 2010 and >2800million tons of GHG emission by 2020. Indian government in Parliament has announced 25% cut in emission level by 2020 taking 2005 as the base year before the Copenhagen meet. It means that the dependence on energy from coal to renewable will receive priority.

Prime Minister's (Dr. Manmohan Singh as prime minister) council on climate change has approved eight missions under India's National Action Plan on Climate Change (NAPCC).

Under the ambit of NAPCC, eight Missions have been initiated to implement the programmes related to mitigation and adaptation.

The missions are:

- National Solar Mission
- National Mission for Enhanced Energy Efficiency in Industry
- National Mission on Sustainable Habitat
- National Water Mission

- National Mission for Sustaining the Himalayan Ecosystem
- National Mission for a „Green India
- National Mission for Sustainable Agriculture
- National Mission on Strategic Knowledge for Climate Change

These eight Missions form the core of National Action Plan representing

Multi-pronged long term and integrated strategy for achieving the goals in the Context of climate change.

The PM's Office has announced that by 2015, India will save about 5% of annual energy consumption & nearly 100 million tons of CO_2 emissions per year. The National Solar Mission has approved enhancement of solar energy generation.

On 21st October 2009, India & China have signed a MOU for collaborating on climate change issues at mutual and international level. India launched 83Kg small observation satellite (IMS-I) in April 2008. On 18th October 2008, Mr.G.Madhavan Nair announced further actions to be taken: India plans to launch two climatc satellites (One 50 Kg micro satellite for climate study and the Second satellite (Remote sensing) will monitor GHG (CO_2, & CH4) emission in polar orbit

The Satellite will contain 2 optical instruments-

(i) a multispectral camera,

(ii) a hyper-spectral camera to operate in visible and near infrared regions.

These actions indicate India's concern and positive voluntary actions.

India's Concern

The present government showed India's comcern before the Paris Agreement. Population-1.3 billion in 2016 to grow ~ 1.6 billion by 2030 or so. India is committed to annual economic growth ~ 7-8%. Indian's Emissions could increase > 3 times by 2030 as coal is now the main energy source. Being a developing nation, and sizable proportion of people remaining below poverty line, India advocated not shifting to renewable energy fully in very near future. Some flexibility must be allowed.

India's Recommendations

India accounts only about 4% of energy –linked CO_2 Emissions since 1990 in the world. Per capita terms emissions stand at about 1.7 metric tonnes in 2014 (10% of the level of USA % about 25% of the level of China. However the emission is increasing and may reach as predicted by IPCC by 2030. India does not accept the forecast of >2°C & higher rise because there will be technological development in coming years that will curb the temperature increase.

India wanted a ten year review cycle at international level (5 year cycle accepted). India feels that not only historical responsibility, but the advanced nations have the most room to make the cuts & strongest impact (Prime Minister-Narendra Modiji).Minister for power, coal and renewable energy, of India advocated that India's contribution to global greenhouse gases emissions was less than 4% with 17% of the world's population, while developed countries contributed a fifth of emissions with just 5% of the world's population. In view of this differential responsibility is to be practiced.

Paris Agreement:

The conference held in Paris from 30 November, 2015 to 12 December, 2015 is termed as Paris Deal. One hundred ninety five countries participated and by the end of 2016, about 174 countries including India, China, European Union, USA, have ratified the agreement.

Key points of Paris Deal:

1. As expected, the agreement commits the world to limiting global warming to well below 2°C. (Try to keep it below 1.5°C.) so that the adverse effect will be manageable. The deal was a "major leap for mankind.
2. Countries are requested to revisit their existing pledges for emission cuts by 2020, (national reviews in 2018) and every five years
3. On the thorny issue of money, the legally-binding part of the agreement requires that developed countries (as defined in 1992) shall continue to provide finance to developing countries to help them adapt to climate change and cut their emissions. (Technology Development/Transfer).

4. The most contentious part - how much money - is in the non-legally binding 'decision text'. Currently, developed countries are obliged to 'mobilize' \$100 bn a year of public and private finance to help developing countries by 2020.
5. The decision says they "intend to continue their existing collective mobilization goal through 2025" and then by 2025 set a new goal "from a floor of \$100 billion"but, crucially, leaves the actual targets, which are not legally binding, for countries to decide for themselves.
6. "Crucially, countries will review (in 2018) their existing commitments before they come into force in 2020.
7. to achieve that goal, governments pledged to stop the rise in heat-trapping greenhouse gas emissions "as soon as possible". (see the graph below)

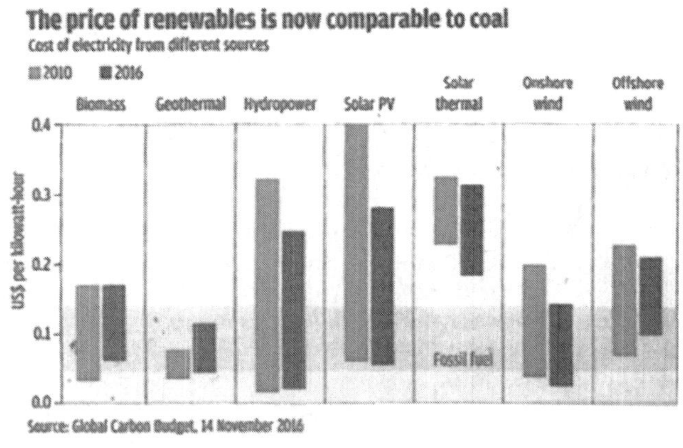

8. by some point after 2050, the agreement says, man-made emissions should be reduced to a level that forests and oceans can absorb. (Hopefully technology will develop)
9. Emissions targets: In order to reach the long-term goal, countries agreed to set national targets for reducing greenhouse gas emissions every five years not ten years as India suggested.
10. More than 180 countries have already submitted targets for the first cycle beginning in 2020.
11. Only developed countries are expected to slash their emissions in absolute terms; developing nations are "encouraged" to do so as their capabilities evolve over time.

Until then, they are expected only to rein in growth of emissions as their economies develop.

Positive Points of Paris Deal

The agreement includes some of the elements of an ambition mechanism such as:

(i) 5-year cycles, emission reduction actions,

(iii) Finance and adaptation,

(ii) Periodic review of global stock,

(iv) Establish a clear long-term 2050 goal to move away from fossil fuels and to renewable energy and sustainable land use

(v) by including a long-term temperature goal of well below 2°C of warming and a reference to a 1.5°°C goal, the agreement sends a strong signal that governments are committed to being in line with science.

(vi) In addition the recognition of the emissions gap and the inclusion of a quantified 2030 gigatonne goal should serve as a basis for the revision of national pledges ahead of 2020.

(vii) The agreement includes a section recognizing "loss and damage" associated with climate-related disasters. Promises to deliver support to vulnerable countries to limit climate impacts and address unavoidable damage. This will help small island nations.

USA will cut GHG emissions by 26% to 28% from 2005 levels by 2025.Building on the U.S. Strategy to Reduce Methane Emissions, the Administration announced a goal to cut methane emissions from the oil and gas sector by 40 to 45 percent from 2012 levels by 2025 and has committed to finalize recently proposed standards for methane emissions from the oil and gas sector in 2016. USA is committing to reduce methane emissions from landfills by 2016. U.S. Clean Power Plan will reduce emissions in the U.S. power sector by 32% by 2030. USA is stepping up the work to phase down super-polluting hydro fluorocarbons (HFCs).EU will cut emissions by 40% from 1990 level by 2030.China will peak its emission cut up to 40% by 2030. China confirmed that it plans to launch in 2017 a national emission trading system covering power generation, steel, cement, and other key industrial sectors, as well as implement a "green dispatch" system to favor low-carbon sources in the electric grid. China is stepping up work to phase down super-polluting hydro fluorocarbons (HFCs).

Indian Commitment: Before and after the Paris Deal, & Action Taken:

1. India's centrepiece a commitment to derive 40% of its electricity from renewables and other low carbon sources. India has installed base of 42, 000 MW of renewable energy by 2012 and the plan was to increase it to 70, 000MW by 2020, & now the commitment is one lakh MW from renewables by 2030 and this is part of the programme of energy security for the country and reduces carbon emissions to the atmosphere.
2. India has plans to add 30 times more solar-powered generation capacity by 2022. Clean coal would remain the most efficient energy source for decades in India considering the huge quantity of coal deposits in the country.
3. India plans for the use of ecofriendly transport fuels to reduce carbon emissions to the atmosphere. About 5 million automobiles run in Indian roads (2016) and India consumes about >1200 million litre of petroleum products (petrol, diesel etc) per day (google search- Wikipedia). In India, ethanol blending (5%) of the diesel etc (BS-III & BS-IV fuels) have been started from 2012-2013 reducing pollutant load and carbon load to the atmosphere. India planned to introduce BS-VI fuel by April, 2017 (has not been done but efforts are made to implement from 2018), which will further reduce the pollutant load and carbon load to the atmosphere (see below). FAME (fatty acid methyl ester products), HVO (hydrogenated vegetable oil), Biofuel, Electricity, hybrids, Hydrogen are the near future fuels for transport. Research in this field is progressing well in many countries including India. Ethanol is usually made from crops (grains), & traditional feed stocks. Ethanol made from cellulosic biomass material is called Bio-ethanol- Biofuel. Cellulosic & Ligno-cellulosic materials, non-agricultural biomass, non feedstock materials, Jatropha oil etc are used to produce Biofuel. (Paribesh Samachar, 2016).

Emission (per km) ↓ Fuels →	**BS-IV**	**BS-VI**
CO_2	130g/km	95g/km for cars
	175g/km	147/km for heavy vehicles
CO	300mg/km	170mg/km
SO2	50ppm	10ppm
NOx	250ppm	80ppm
PM	25mg/km	05mg/km

CNG (Compressed Natural Gas) Buses and vehicles ~10, 000 numbers now run in Delhi Roads- to reduce pollution & GHG emission. LPG (Liquified Petroleum Gas- mainly propane & Butane) is also being experimented (google.com search: Wikipedia).

Research in this field is progressing well in many countries including India.

Indian Parliament has approved that thirty three percent of land area with forest cover, twenty percent of urban centres with tree cover, creating number of protected areas (PAs) & to increase carbon sink-sequestration potential in the country.

Conclusion:

In 2016, 'the Earth Day pledge was 'Trees for the Earth'. Over the next five years, as Earth Day moves closer to its 50th anniversary, (started 1970). UNO is calling on you to help us achieve one of our most ambitious goals yet —The plan is to plant 7.8 billion trees and we're starting now. Trees will be the first of five major goals we are undertaking in honor of the five-year countdown to our 50th anniversary. On their own and together, these initiatives will make a significant and measurable impact on the Earth and will serve as the foundation of a cleaner, healthier and more sustainable planet for all'.

Policy to Grow Trees?

Trees help combat climate change. They absorb excess and harmful CO_2 from our atmosphere. In fact, in a single year, an acre of mature trees absorbs the same amount of CO_2 produced by driving the average car 26, 000 miles. Trees help us breathe clean air. Trees absorb odors and pollutant gases (nitrogen oxides, ammonia, sulfur dioxide and ozone) and filter particulates out of the air by trapping them on their leaves and bark. Trees help communities. Trees help communities achieve long-term economic and environmental sustainability and provide food, energy and income. (Google search-Earth day, Wikipedia). In 2016, on the World Environment Day, UNO

Decided to encourage nations to plant about 8 billion trees and the programme was a great success.

Further Reading

Dash, M.C. and S.P.Dash.2009. Fundamentals of Ecology, McGraw-Hill Publications, New Delhi, 562 pages,

Down To Earth, 2017. State of India's Environment, Centre for Science & Environment, New Delhi, 226 pages,

Down To Earth, 2016. State of India's Environment, Centre for Science & Environment, New Delhi, 208 pages

Down To Earth, 2015.State of India's Environment, Centre for Science & Environment, New Delhi, 208 pages,

Down To Earth, 2009.Climate Change: Politics and Facts, Centre for Science & Environment, New Delhi, 125 pages.

Google.com.search: Impact of Climate Change on Agriculture, Wikipedia, 2017,

Horn Bill, BNHS, Bombay, 2015-2017 issues.

IPCC Assessment Reports on Climate Change, 1990 (1st Report), 1992 (supplement), 1995 (2nd Report), 2001 (Third Report), 2007 (4th Report), 2014 (5th Report). UNEP, New York/Geneva,

Paribesh Samachar, 2016. State Pollution Control Board, Odisha.

Statistics Related to Climate Change – India 2015: Government of India, Ministry of Statistics and Programme Implementation, Central Statistics Office, Social Statistics Division, New Delhi

The Climate Reality Project, India.2016. Understanding Climate Change, UNESCO, New Delhi Office, 4 pages.

www.downto earth.org.in

15

Climate Change and Paris Deal: A Review

Introduction

On 29th April, 2016, Mr. Ashish Mohapatra of Calgary, Canada visited us at Bhubaneswar .The province of Alberta in Canada is a very cold place for about five-six months in a year and winter is full of snow. Calgary is a modern and beautiful city in Alberta where the per capita income and per capita number of cars is one of the highest among the north-American cities. He was narrating the effect of global warming and climate change and opined there was a very mild winter with no snow accumulation in Calgary in 2015 and the summer temperature was about 35-37°C, about 5-7°C higher than the usual summer temperature.

We have observed very harsh summer throughout India in 2016.Chennai experienced unprecedented heavy rains in November-December, 2015, flooding the city and causing disastrous situation for the people. This year, the shortage of drinking water in many parts of the country including Odisha is severe and has brought misery to the people.

Scientists involved in studying and preparation of the IPCC (Intergovernmental Panel on Climate Change, UNO) climate change reports (IPCC, 2007, and 2014) have conclusively proved that climate change is real and its effects are disastrous. The adverse effects are now visible in many parts of the world, especially in south-east Asia including India in the form of severe cyclone, flooding in some areas, drought in some areas, reduced crop production etc. Hibernation period in many vertebrate species is shortened affecting their metabolism and life cycles. Coral reefs are bleaching (BNHS, 2016) and seriously affecting fish diversity and life cycles. This situation has been realised world over since 1960s and efforts have been made by UNO (Table 1) to arrive at principles of taking urgent action by the different nations of the world.

Table 1. Road map to cooler world- UNO initiatives

- 1972 Stockholm – Man & His Environment- UNEP formed
- 1979-Montreal – Climate Change
- 1983-1987-Brundland Commission Report- Our Common Future
- 1989- Intergovernmental Panel of Climate Change was formed
- 1992 Rio de Janeiro – Earth Summit- Sustainable Development, Agenda-21-Biodiversity, Climate Change etc
- 1997 Kyoto Protocol – Climate Change-Carbon Trading
- 2002 Johannesburg- Review
- 2007 Bali Road Map- Climate Change
- 2008-Pozanan Conference- Climate Change
- 2009 Bonn, Germany-Climate Change
- 2009 Copenhagen- Climate Change
- 2010 Cancun, Mexico- Climate Change
- 2011 Johannesburg- Climate Change & Sustainable Development
- 2012 Durban- Climate Change
- 2012 Doha- Climate Change
- 2012 Rio de Janeiro- Climate Change, Sustainable Development, - The Future we want- Common but Differential Responsibility
- 2015 Paris, France- 195 countries have participated (Nov.30-Dec.12)

UNO is persuading nations to take urgent actions on mitigation measures and adoption of different methods to address the reduction of effects of climate change. This matter was discussed in Paris during 30 November to 12 December, 2015. This paper discusses the global warming and climate change problem addressed in on Paris Deal.

International Land-mark conferences on the problems of climate change:

Kyoto Protocol-1997: (i) All countries were grouped into two categories- (a) Those consuming energy > than world average (industrialized countries termed Annex-I countries), & the rest were classified as Annex-II countries, (ii) Annex-I countries were to bring down their GHG emissions by 5.2% from the base level of 1990 keeping in mind their economic development, Population growth etc in mind. (iii) If for reasons they are not in a position to bring down below 1990 level, they were allowed carbon trading with Annex-II countries. (iv) The time period allowed GHG emission reduction was from January 2000 to 2012, later extended to 2015. (v) The Clean Development Mechanism was elaborated. Few developed countries namely USA and Australia did not sign this document. They wanted that the large economies like China, India, Brazil, Mexico, and South Africa should take steps to reduce GHG emission although these countries were listed as Annex-II.

Bali Road Map-2007: 1.Adoption Fund was created ($50 billion), 2. Technology Transfer was made simple, (3) USA & Australia agreed to sign the post 2012 agreement, (4) the message was 'The world cannot afford to wait'. Poznan, Poland-2008: 1.Agreed to extend the Kyoto protocol- carbon trading up to 2015. (2) UNFCCC pointed out that CDM process does not properly account for Building sector, (3) The adoption fund was thought to be meager. Policy shift by USA-2009- They agreed that it is an environment issue, health issue, and economic issue, energy use issue & climate change is real. (Mr.Barack Obama became President).

Copenhagen-2009 (i) No specific emission targets set for developed countries, (ii) Consensus was to keep temp. rise below 2°C by 2050AD, (iii) Emphasis was on mobilising finance to support reforestation, especially in developing countries. (iv) International monitoring was not accepted (v) Political accord was made but no legal binding, (vi)

Nations agreed to scale up public finance for Green House Gas (GHG) reduction. (Dash, 2004, Down to Earth, 2009).

The key issue is to reduce CO_2e emissions to the atmosphere so that the effect of global warming can be controlled. CO_2e emissions are linked to the development. The energy source from renewables require costly technology to harness energy from sun, wind power, geothermal, biomass gasification etc. Hence Technology transfer with affordable cost is the key issue. (Down to Earth, 2015)

Table 2. Green House Gases (GHG) :
(i) Carbon Dioxide (CO_2) : Source (i) Automobiles, (ii) Industrials, (iii) Anthropogenic: Increased from 280 ppm to about 383ppm (2007), ~395ppm in 2012, ~ 403ppm in 2015. Absorbs Heat at three bands (nm) : 2.3 to 3.1, 4.1-4.5, 13-18μ, Residence time in the atmosphere (RT) : > 30 years to centuries, GWP=1 (63% 0f Total GHG)
(ii) Methane (CH4) : Source (i) Organic Matter decay under anaerobic conditions, (ii) Agriculture fields. (iii) Ruminant enteric fermentation- Increasing. Absorbs heat at two bands (nm) : 3.2 & 7.6μ, RT:14 years, GWP=25 (4% of Total GHG)
(iii) Nitrous Oxides (N2O) : Source (i) Automobiles, (ii) Industries, (iii) Agriculture & Live stock: Increasing. Absorbs heat at band (nm,) : About 7.66μ, RT:114 Yrs, GWP-298 (< 9%) (iv) Water Vapor: Source (i) Evaporation, (ii) Transpiration: Evaporation increasing. Absorbs heat at bands (nm,) :1 to 8μ (of Total GHG) 23-25%. (v) HydroFlurocarbons (HFC) / Perflurocarbons (PFCs) /Sulphur hexafluoride (SF6) : (1-2% of Total GHG) - Montreal Protocol-Shifting from HClFCs ——!HFC——!Synthetic substances. Source: (i) Refrigerants, (ii) Foam, (iii) Fire extinguishers, (iv) Aerosol products (v) Some mineral based industries- Increasing, Absorbs heat at bands (nm) : 7-12μ, RT:1.4 to 270, yrs, PFC-10, 000-50, 000 Yrs, GWP=1000-23900, (1%), Ozone layer depletion.

Source: IPPC, 2007 (4th Assessment Report of IPCC), CUP, Cambridge.

Figure 1 shows the expected atmospheric temperature rise as per IPPC, if aggressive mitigation measures are not attempted immediately.

The threshold is estimated as a temperature rise of 2°C above pre-industrial levels, and on current emissions trajectories, we are heading for a rise of about 5°C without significant mitigation between 2050-2100AD.That may not sound like much, but the temperature difference between today's world & 1900AD is 0.83°C and the difference from

Graph 2.6: Expected temperature rise till 2100

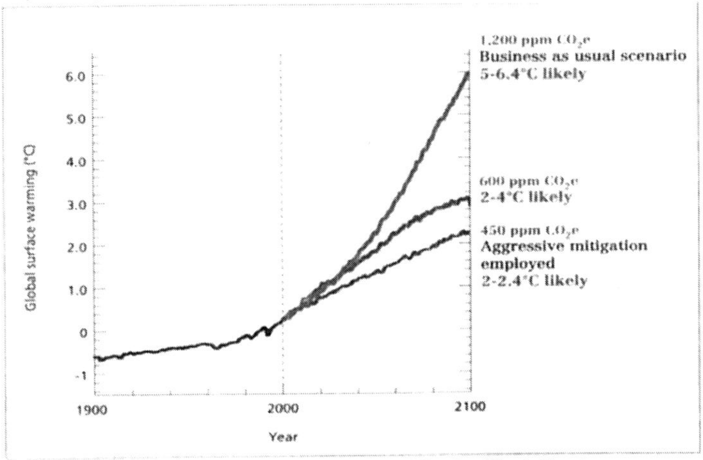

Source: Fourth Assessment Report, 2007, Intergovernmental Panel on Climate Change

Expected Temperature Rise: IPCC

the last ice age (about 12 thousand years before) is about 5°C, so seemingly small changes in temperature can mean big differences for the Earth.

Effects of Global warming already observed:

The following photographs/diagrams exhibit the recently observed effects of global warming and climate change. (Figures 2-10)

Vulnerable Points

Map 2.1: Earth's tipping points (potential)

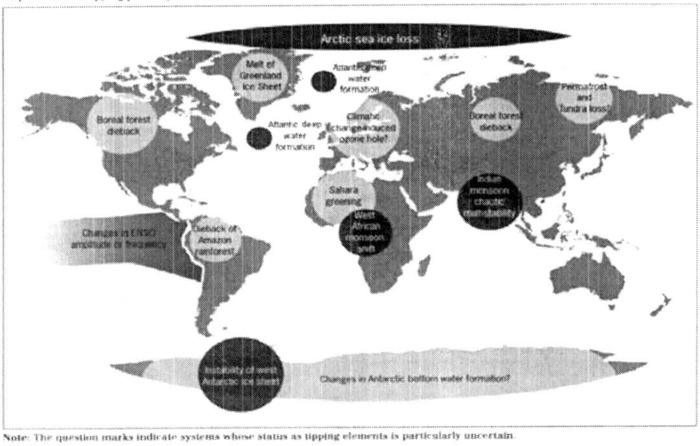

Note: The question marks indicate systems whose status as tipping elements is particularly uncertain.

Source: Richardson, Katherine *et al* 2009, *Synthesis Report: Climate Change: Global Risks, Challenges and Decisions, Copenhagen 2009, March 10-12*, University of Copenhagen, June, p 14

Vulnerable hot spots

Map 5.2: The IPCC's Fourth Assessment Report (AR4) clearly identifies deltas and megadeltas as hotspots of vulnerability

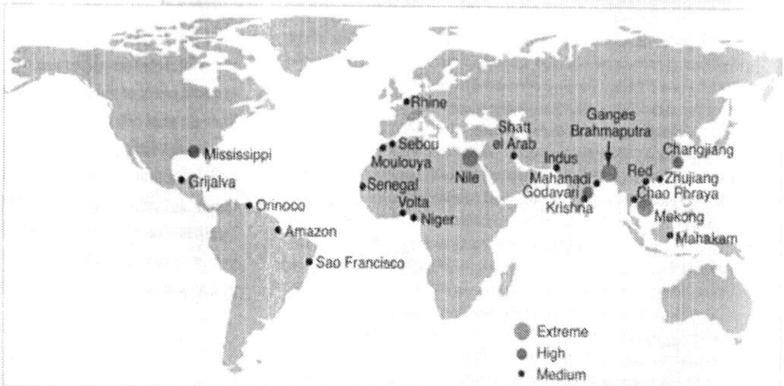

Note: Relative vulnerability of coastal deltas as shown by the indicative population potentially displaced by current sea level trends to 2050 (Extreme = >1 million; High = 1 million to 50,000; Medium = 50,000 to 5,000; following Ericson et al, 2006).

Chennai Flood-2015

(taken from Down to Earth magazine)

- An abnormal rise in temperature in sea surface water in the equatorial pacific ocean, called El Nino, affects the atmosphere temperature and monsoon current in South-east Asia, especially Indian subcontinent.
- 0.5°C over the normal sea surface temperature indicates El Nino effect has set in.
  - (i) 0.5 to 0.9°C rise is caused by weak El Nino,
  - (ii) a rise 1.0 to 1.5°C is a result of moderate El Nino and

(iii) > 1.5°C rise over the normal sea surface temperature is classified as strong El Nino.

In 2015, a strong El Nino prevailed from July onwards causing deficient rain fall and caused drought in some parts of India and very high rain fall in Some other parts, like Tamil Nadu causing heavy flood. This phenomenon happens naturally and intensely aided by Heat Island effect created by Humans in the large scale urbanization (concrete houses, asphalt roads, loss of water bodies, unusual traffic density created by us, refrigeration, ACs etc produce high GHG) leading to warming. (Down to Earth, 2015).

Erratic Rainfall-High variability

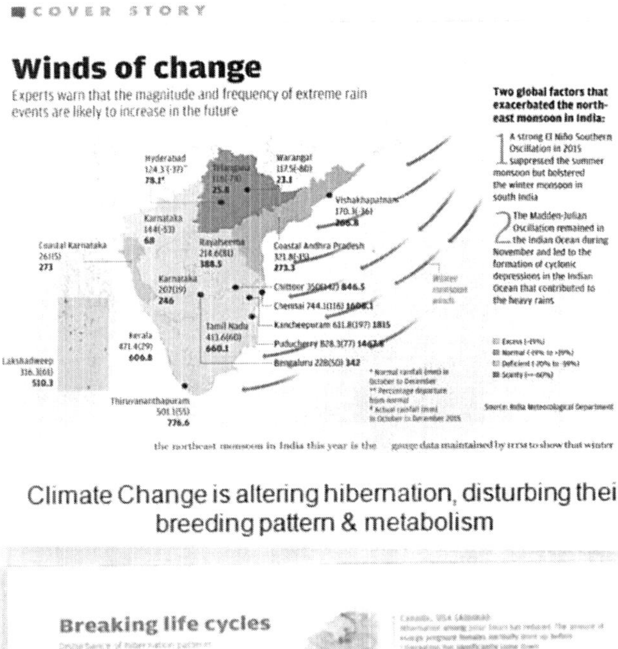

Climate Change is altering hibernation, disturbing their breeding pattern & metabolism

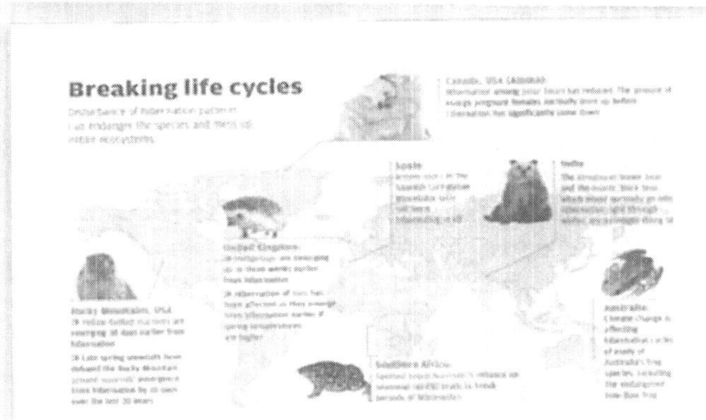

(SPAN, 2015, 2016)

Coral Reefs are leaching with tremendous loss to biodiversity

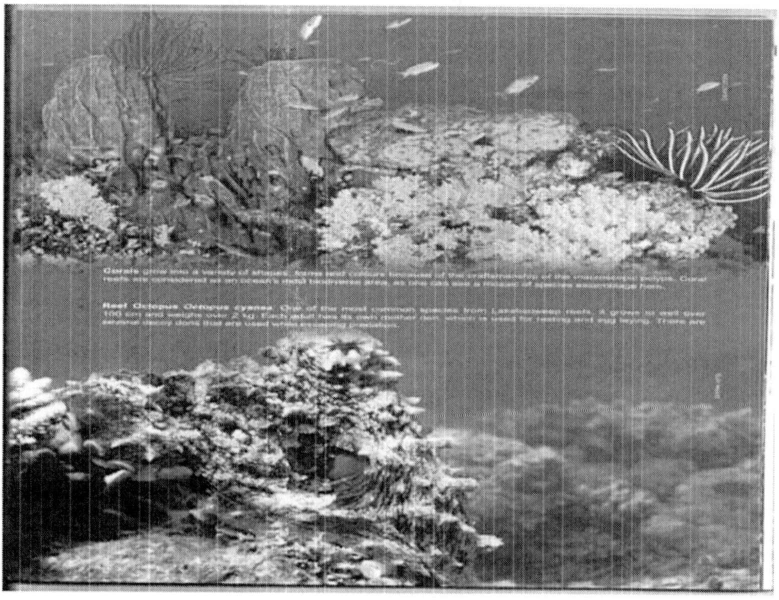

(Taken from BNHS, Horn Bill 2016)

Himalayan Glaciers are melting (Down to Earth 2014)

Paris Deal

Earth's average atmospheric temperature in 2015 is ~15°C, which is on the average 0.83°C higher than the average temperature which prevailed in1900 AD. IF emissions keep rising unchecked, temperature could rise an extra 4°C or higher by 2050. If emissions cut immediately & the rise can be limited to ~1.5°C. Globally agreed threshold temperature rise is 2°C and beyond which the earth will experience conditions very difficult for Survival of Human Beings & Biodiversity.

The present source of energy use is the key: Efforts are made to shift from coal and other fossil fuels to alternate fuel.

Table 3. COAL CONSUMPTION (MT per yr- 2014 data) % in Electricity Mix.

Country	Total	per Capita (T per yr)	Coal	Gas	Total Fossil
China	1960	1.4	72	2	74
USA	453	1.5	38	30	68
EU	270	0.5	36	15	51
India	360	0.3	70	6	76
Japan	127	1	30	43	73

GHG Emission Sources

Total GHG Emissions ~ 49 Gigatonnes of CO_2e by 2010 & ~ 54 Giga tonnes in 2014., with 2.2% increase annually.

Electricity & Heat Production	25%
Agriculture & Forestry	24%
Industry	21 %
Transport	14%
Buildings	06.4%
Miscellaneous	09.6%

Source: Fifth Assessment Report of IPCC (2014-as Reported by'Down to Earth'-16-30 November 2014, pages 54-55 pages)

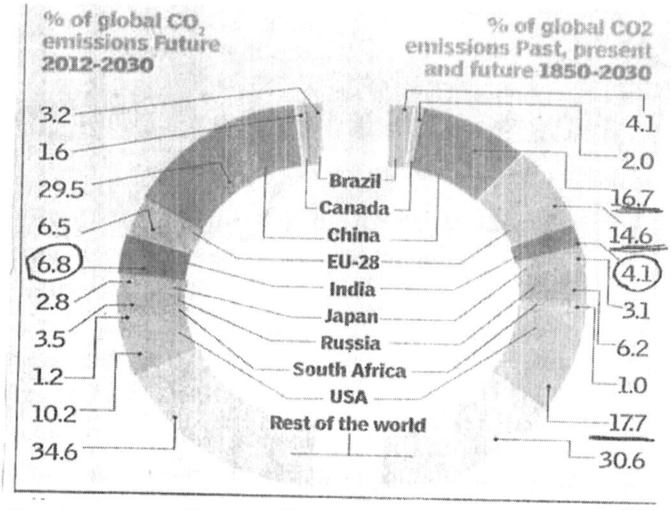

The following diagram (Figure 12) shows percentage of global CO_2e emission from 1850 to 2030 and expected emissions from 2012-2030. India's per capita emission on the average may go up to 4.1 tonnes per year by 2030 and it is predicted that the emissions during 2012 to 2030 may go upto 6.8 tonnes per year in comparison to 29.5 tonnes in China, 17.7 tonnes in USA and 14.6 tonnes in European Union. (IPCC, 2014)

The total emissions per year at present amount to about 54Gt and the total emissions will amount to about 748.2Giga tonnes of CO_2e by 2030 AD resulting an accumulation of 2900 Gt in the atmosphere (Figure 13)

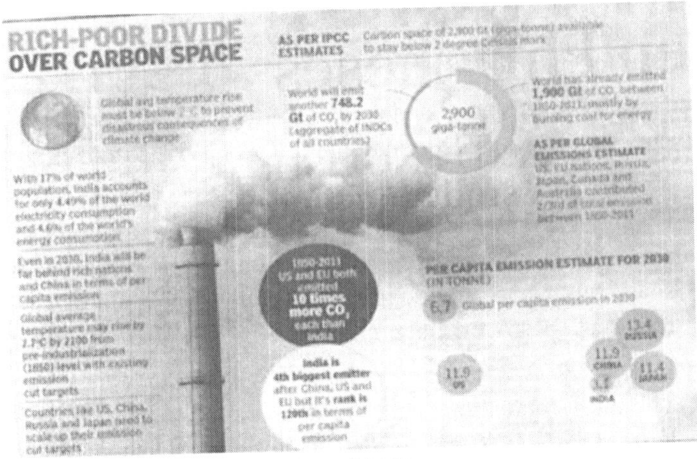

(Times of India Publication, 2015)

India's Concern

Population-1.25 billion in 2015 to grow ~ 1.6 billion by 2030. India is committed to annual economic growth ~ 7-8%. Indian's Emissions could increase > 3 times by 2030 as coal is now the main energy source.

INDIA's Recommendations

India accounts only ~4% of energy –linked CO_2 Emissions since 1990, .Per capita terms emissions stand at ~1.7 metric tonnes in 2014 (10% of the level of USA % ~25% of the level of China. However the emission is increasing and may reach as predicted by IPCC by 2030. India does not accept the forecast of >2.°C & higher rise because there will be technological development in coming years that will curb the temperature increase. India wanted a ten year review cycle of review at international level (5 year cycle accepted). India feels that not only historical responsibility, but the advanced nations have the most room to make the cuts & strongest impact (Prime Minister-Narendra Modiji).Minister for power, coal and renewable energy, Piyush Goyal, said India's contribution to global greenhouse gases emissions was just 2.5-3% with 17% of the world's population, while developed countries contributed a fifth of emissions with just 5% of the world's population. In view of this differential responsibility is to be practiced.

Key points of Paris Deal

1. As expected, the agreement commits the world to limiting global warming to well below 2°C. (try to keep it below 1.5°C.) so that

the adverse effect will be manageable. The deal was a "major leap for mankind.

2. Countries are requested to revisit their existing pledges for emission cuts by 2020, (national reviews in 2018) and every five years
3. On the thorny issue of money, the legally-binding part of the agreement requires that developed countries (as defined in 1992) shall continue to provide finance to developing countries to help them adapt to climate change and cut their emissions. (Technology Development/Transfer).
4. The most contentious part - how much money - is in the non-legally binding 'decision text'. Currently, developed countries are obliged to 'mobilize' $100 bn a year of public and private finance to help developing countries by 2020.
5. The decision says they "intend to continue their existing collective mobilization goal through 2025" and then by 2025 set a new goal "from a floor of $100 billion"but, crucially, leaves the actual targets, which are not legally binding, for countries to decide for themselves.
6. "Crucially, countries will review (in 2018) their existing commitments before they come into force in 2020.
7. to achieve that goal, governments pledged to stop the rise in heat-trapping greenhouse gas emissions "as soon as possible".
8. by some point after 2050, the agreement says, man-made emissions should be reduced to a level that forests and oceans can absorb. (Hopefully technology will develop)
9. Emissions targets: In order to reach the long-term goal, countries agreed to set national targets for reducing greenhouse gas emissions every five years not ten years as India suggested.
10. More than 180 countries have already submitted targets for the first cycle beginning in 2020.
11. Only developed countries are expected to slash their emissions in absolute terms; developing nations are "encouraged" to do so as their capabilities evolve over time.

Until then, they are expected only to rein in growth of emissions as their economies develop.

Positive Points

The agreement includes some of the elements of an ambition mechanism such as

(i) 5-year cycles, emission reduction actions,

(ii) Finance and adaptation,

(iii) Periodic review of global stock,

(iv) Establish a clear long-term 2050 goal to move away from fossil fuels and to renewable energy and sustainable land use

(v) by including a long-term temperature goal of well below 2°C of warming and a reference to a 1.5°°C goal, the agreement sends a strong signal that governments are committed to being in line with science.

(vi) In addition the recognition of the emissions gap and the inclusion of a quantified 2030 gigatonne goal should serve as a basis for the revision of national pledges ahead of 2020.

(vii) The agreement includes a section recognizing "loss and damage" associated with climate-related disasters. Promises to deliver support to vulnerable countries to limit climate impacts and address unavoidable damage. This will help small island nations. USA will cut GHG emissions by 26% to 28% from 2005 levels by 2025.Building on the U.S. Strategy to Reduce Methane Emissions, the Administration announced a goal to cut methane emissions from the oil and gas sector by 40 to 45 percent from 2012 levels by 2025 and has committed to finalize recently proposed standards for methane emissions from the oil and gas sector in 2016. USA is committing to reduce methane emissions from landfills by 2016. U.S. Clean Power Plan will reduce emissions in the U.S. power sector by 32% by 2030. USA is stepping up the work to phase down super-polluting hydro fluorocarbons (HFCs).EU will cut emissions by 40% from 1990 level by 2030.China will peak its emission cut up to 40% by 2030. China confirmed that it plans to launch in 2017 a national emission trading system covering power generation, steel, cement, and other key industrial sectors, as well as implement a "green dispatch" system to favor low-carbon sources in the electric grid. China is stepping up work to phase down super-polluting hydro fluorocarbons (HFCs).

Indian Commitment: Before and after the Paris Deal, & Action Taken

1. India's centrepiece a commitment to derive 40% of its electricity from renewables and other low carbon sources. India has installed base of 42, 000 MW of renewable energy by 2012 and the plan was to increase it to 70, 000MW by 2020, & now the commitment is one lakh MW from renewables by 2030 and this is part of the programme of energy security for the country and reduces carbon emissions to the atmosphere.

2. India has plans to add 30 times more solar-powered generation capacity by 2022. Clean coal would remain the most efficient energy source for decades in India considering the huge quantity of coal deposits in the country..

3. India plans for the use of ecofriendly transport fuels to reduce carbon emissions to the atmosphere. About 5 million automobiles run in Indian roads (2016) and India consumes about >1200 million litre of petroleum products (petrol, diesel etc) per day (google search- Wikipedia). In India, ethanol blending (5%) of the diesel etc (BS-III & BS-IV fuels) have been started from 2012-2013 reducing pollutant load and carbon load to the atmosphere.

•About 5 million automobiles run in Indian roads (2016) and India consumes about >1200 million litre of petroleum products (petrol, & diesel) per day (google search- Wikipedia). India plans to introduce BS-VI fuel by April, 2017 which will further reduce the pollutant load and carbon load to the atmosphere. FAME (fatty acid methyl ester products), HVO (hydrogenated vegetable oil), Biofuel, Electricity, hybrids, Hydrogen are the near future fuels for transport. Research in this field is progressing well in many countries including India. Ethanol is usually made from crops (grains), & traditional feed stocks. Ethanol made from cellulosic biomass material is called Bio-ethanol- Biofuel. Cellulosic & Ligno-cellulosic materials, non-agricultural biomass, non feedstock materials, Jatropha oil etc are used to produce Biofuel. (Parivesh, 2016).

Emission (per km) ↓ Fuels → BS-IV BS-VI

CO_2	130g/km	95g/km for cars
	175g/km	147/km for heavy vehicles
CO	300mg/km	170mg/km
SO2	50ppm	10ppm
NOx	250ppm	80ppm
PM	25mg/km	05mg/km

CNG (Compressed Natural Gas) Buses and vehicles ~10, 000 numbers now run in Delhi Roads- to reduce pollution & GHG emission. LPG (Liquified Petroleum Gas- mainly propane & Butane) is also being experimented (google.com search: Wikipedia).

Research in this field is progressing well in many countries including India.

Indian Parliament has approved that thirty three percent of land area with forest cover, twenty percent of urban centres with tree cover, creating number of protected areas (PAs) & to increase carbon sink-sequestration potential in the country.

Conclusion:

This year's (2016), 'the Earth Day pledge was 'Trees for the Earth'. Over the next five years, as Earth Day moves closer to its 50th anniversary, (started 1970). UNO is calling on you to help us achieve one of our most ambitious goals yet —The plan is to plant 7.8 billion trees and we're starting now. Trees will be the first of five major goals we are undertaking in honor of the five-year countdown to our 50th anniversary. On their own and together, these initiatives will make a significant and measurable impact on the Earth and will serve as the foundation of a cleaner, healthier and more sustainable planet for all'.

Why Grow Trees?

Trees help combat climate change. They absorb excess and harmful CO_2 from our atmosphere. In fact, in a single year, an acre of mature trees absorbs the same amount of CO_2 produced by driving the average car 26, 000 miles. Trees help us breathe clean air. Trees absorb odors and pollutant gases (nitrogen oxides, ammonia, sulfur dioxide and ozone) and filter particulates out of the air by trapping them on their leaves and bark. Trees help communities. Trees help communities achieve long-term economic and environmental sustainability and provide food, energy and income. (Google search-Earth day,

Wikipedia). Environmental discipline and population control are key issues for a sustainable society. Ecologists/Biologists can work on carbon flow in the ecosystem and suggest adoption of carbon sink efficiency using the following model.

Carbon Flow Model in the Ecosystem

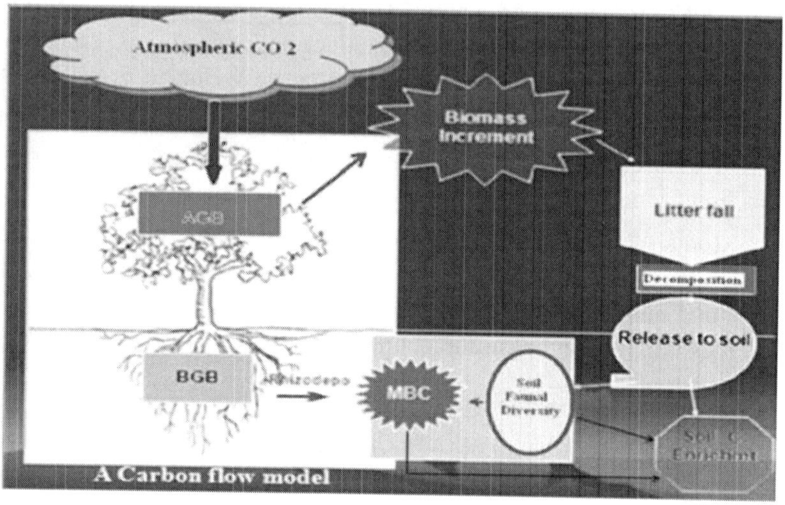

References:

Horn Bill. 2016. Bombay Natural History Society, (BNHS), Mumbai.

Centre for Science and Environment, 2009. Climate Change: Politics and Facts. 125 p., CSE, New Delhi, 2009,

Down to Earth-2014. Climate Change, 16-30 November 2014, pages 54-55 pages, CSE, New Delhi, 2014..

Down to Earth, 2015.A Down to Earth Annual State of India's Environment, 2015.Centre for Science and Environment, New Delhi 2015.

Down to Earth, 2016. August issue. Centre for Science and Environment, New Delhi 2016.

Google search 2016-Earth day, Euro-VI, Climate Change, Wikipedia,

IPCC (UNEP). 2007. Climate Change: A Report of the Intergovernmental Panel on Climate Change, Cambridge University Press, U.K. & New York, 4 volumes.

IPCC (UNEP) 2014. Climate Change, Vol.5 (Quoted by Down to Earth, 2014)

Times of India, 2015. December Issues.

16

A Comparison of Industrial Green House Gas Emissions (GHG) and Carbon Sink Potential of Forest Vegetation in Orissa, in the Context of Climate Change

INTRODUCTION

Climate change is popularly understood to mean the effect of global warming. Climate flip-flop is now very common. During the last few hundred years of human activities, particularly, industrialization, urbanization and ecosystem fragmentation have caused serious damage to the planet culminating in climate flip-flop that every nation is now concerned. Energy source and its use in farm, industry and service sectors is the key to understand its impact on climate chage. Earth's atmospheric temperature is rising causing global warming. Scientific evidences indicate that the anthropogenic activities are aiding to the global warming (UNEP, 2007, CSE, 2009). Climate change may be due to many factors, i.e. natural internal processes in the planet, external factors, but persistent anthropogenic activities from industrialization, use of fossil fuel as main energy source, transport (an estimate shows that 650 million automobiles run in the world), land use etc bring changes in the atmospheric composition.

Taking the late 17th century as base year, scientists tell us that the carbon dioxide level was around 280ppm During the last 100 years or so, it has increased to 387ppm (1870 to 1986-240ppm to 340ppm, 1986-2005:340 to 383ppm and now at about 387 ppm) and at this rate (2 ppm per year) of increase, the CO_2 level will reach about 550ppm by 2050 and 700-1200 ppm by 2100 AD. Besides, the concentration of other green house gases like methane, chlorofluorocarbons, perflurocarbons, oxides of nitrogen and some other gases, whose resident time in the atmosphere is very long has increased.

This paper estimates and reviews the green house gas emission potential of industries of Orissa, a fast growing state for coal based power plants, iron and steel, aluminium and cement industries and the carbon sequestration potential of forests and related issues. The estimation is based on the published data on industrial production, forest cover, and CO_2 absorption capacity of forests in India. The carbon sequestration potential of forests has been collected from the publications of MOEF, New Delhi (FSI, 2009) and has also been estimated by estimating the net primary productivity in different world sites (Lieth, 1972 and 1977; Dash and Dash, 2009). A reference has been made to the situation in Orissa as industrialization has been taken as a thrust area of development currently due to rich mineral deposits in the state. (The State of Environment: Orissa, Dash (Ed), 2006; Dash, 2010; Paribesh Samachar, 2010).

Green House Gas emissions from industries

Industrialization has grown manifold in India after 1947. An estimate of GHG emissions and the basis of estimations are given in Table 1. India annually burns

- 300 million tonnes of coal in power plants and generates equal amounts of CO_2 from this source.

There is increased need of energy in the farm, industrial and service sectors as the

Received : 15.06.2010
Revised : 21.07.2010
16.08.2010

Table 1: The basis of estimations of GHG emissions from energy sector and important industrial sectors in the country with the capacity of production

Energy sector: Thermal power	1MWH generation requires about 17 tons of Indian Coal, which contains 32% C. 1MWH power generates about 0.79 t CO_2 .Hence 1000MWH, generates about 6.92 million tons CO_2 per year. India generates ~1.00,000 MW thermal power annually and emits about 692 million tonnes of carbon dioxide.
Industrial process sector	Iron and steel production: 1 tonne production generates about 3-3.5 tons of CO_2 from coal and coke. India produced ~ 44 million tonnes of iron and steel in 2006 with about 10% annual growth. The estimate of current production is about 65 million tonnes of iron and steel per year. (65 x 3.5=227.5 million tonnes of CO_2). Future projection is > 100 million tonnes by 2020.
FERRO- CHROME	1 tonne production generates about 2.1 tonne of CO_2. The current Ferro-Chrome demand is estimated as 8 million tonnes (8x2.1 = 16.8 million tones of CO2 generation). (Taking the values of 84% C and 50% C in metallurgical useable grade coke and coal respectively).
Alumina and aluminium	1 tonne alumina production using about 4 tons bausite ore and 1 tonne aluminium productionusing about 2 tons of alumina generate about 21.66 ton CO_2. India has the capacity of ~ 4 million tonne aluminum production per year. (CO_2 emission amounts to 86.64 million tonnes)
Cement	1 ton production generates about 0.9 ton CO_2. In 2003, India produced about 117 million tonnes of cement. At present, India has ~ 200 million tonne annual production capacity. (180 million tonnes of CO_2 emission).

population continues to grow and to meet the developmental needs. Modern way and standard of living demand huge energy from unsustainable consumerism, transport facilities and increased construction work and other aspects of human use. India has huge reserves of thermal grade coal and therefore states such as Orissa with abundant coal reserves attract investors to set up thermal power plants if fresh water is available. Coal, therefore occupies important position in power sector. As two major rivers such as Mahanadi and Brahmani flow in Orissa, and the coal mining areas of Angul-Talcher and Ib valley are situated respectively near the rivers Brahmani and Mahanadi, many thermal power plants have been established and are in the pipeline in these areas and likely to be set up there. There is huge pressure on the rivers for fresh water also.

Energy is the key for survival, continuance and for sustainable development.

In 1947 India had electricity generating capacity of only 1470 MW with about 49% load factor. At present India has the capacity to generate >1, 53, 694.09 MW (105 times growth) with about 80% load factor. In spite of this increase, India is not able to meet the demand of energy from the farm, industry and service sectors. Hence the situation has become critical. In the 10th and 11th plans, additional 27, 283 MW and 78, 700 MW could be achieved. It has been estimated that by 2016, the energy production capacity will reach 2, 15, 804 MW. However, per capita energy consumption in India is only about 500 units per year compared to about 2600 units at international level in industrially developed countries.

India's energy source

About 70% of our energy source (1, 53, 694.09 MW) is fossil fuels, especially coal and lignite. The distribution is as follows: (Fatesingh, 2010)

- 81, 355.88 MW-coal based,
- 16, 822.85 MW-Gas based,
- 1199.75 MW diesel based,

Thermal power plants are not environment friendly as these plants generate huge amount of green house gases and pollutants such as SO_2, NOx, Cox, SPM, Fly and bottom ash, metallic dust, etc etc. About

120 million tons of fly ash is generated annually in India (Jain, 2010). The utilization is only 25%-30% in brick making, road construction, forest soil fertilization, and in agriculture etc (Jain, 2010). Hence it is a huge problem. One estimate (Dash, 2007; Paribesh Samachar, 2010) shows that the pollution load from Indian coal based thermal power generation amounts to:

- ~18.96 tonness of CO_2 per day .MW
- >136 kg of SO_2 per day.MW
- >7 tons of fly ash per day.MW

India generates about 1, 00, 000 MW power from thermal source; hence generates about 692 million tonnes of CO_2 per year. From iron and steel, aluminium, and cement industries, the annual GHG emissions may account for 494 million tones (Table 1).

GHG emissions from Orissa

Table 2 and 3 give estimation of GHG emissions from major industries in Orissa.

Three categories of thermal power plants (TPPs), depending upon the type of fuel used operate in the state, i.e. category-I, fuel is only coal either pulverized or sized. The capacity of power generation in category I plants is 7331MW.In category-II, fuel may be low grade coal, washery rejects, and char. These TPPs use Atmospheric Fluidized Bed Combustion (AFBC) or Circulating Fluidized Bed Combustion (CFBC) boilers and these plants are usually small (usual capacity < 30MW). Category-III- TPPs use waste heat of some other processes to generate steam and the plants are usually of low capacity.

The Sink potential

Forest and tree cover of the country was 78.37 million ha in 2007 (FSI, 2009), which was 23.84% of the geographical area Angul-Denkanal area currently account for 61.33% of total thermal power generation and this includes 2.82% tree cover. Table 4 gives the forest and tree cover of the country in 2007 (FSI, 2009).

Forest Survey of India Report, 2009 states that over the last two decades, the national forestry legislation and policies have transformed India's forests into a significant net sink of GHG. The carbon stocks stored in country's forests and tree cover have been estimated as 6,

District/Area	Name	Capacity	Category
Angul-		4,666 MW	
Denkanal	TSTPS, Deepsikha, Kaniha	3000	I
	NALCO Ltd, CPP	960	I
	TTPS, Talcher	460	I
	Bhusan Steel Ltd, Denkanal	110	I
	Nava Bharat Ventures Ltd, Denkanal	094	I
	Bindal Sponge Iron Ltd, Talcher	004	II
	Bhusan Steel Ltd, Denkanal	030	II
	Raja Steel Ltd, Denkanal	004	II
Jharsuguda-		1899.5MW	
Sambalpur	Vedanta Aluminium, Jharsuguda	945	I
	OPGC Ltd, Jharsuguda	420	I
	Hindalco Industries Ltd, Sambalpur	367.5	I
	Bhusan Power and Steel Ltd, Sambalpur	116	I
	SMC Power Generation (P) Ltd, Hirma	17	II
	Eastern Steel and Power Ltd, Jharsuguda	04	II
	Bhusan Power and Steel Ltd, Rengali	60	II
	Shams DRI Power Ltd, Pandaloi	10	II

District/Area	Name	Capacity	Category
	Rathi Udyog Ltd, Chipilima	08	II
	Aryan Ispat and Power Ltd, Rengali	06	II
Sundergarh-		242MW	
Rourkela	NTPC-SAIL Power Ltd, Rourkela	120	I
	RSP, Rourkela	100	I
	Sri Mahavir Ferro Alloys Ltd, Kalunga	08	II
	OCL India, Ltd, Rajgangpur	06	II
	Bhasker Steel and Ferro Alloys Ltd, Bonai	04	II
	Shrishti Ispat Ltd, Budakata	04	II
Jajpur	JSL Ltd, Duburi, Jajpur	250MW	I
Cuttack-Jagatsinghpur		329MW	
	IFFCO Ltd., Paradeep	110	I
	IMFA Ltd, Choudwar	108	I
	Aarti Steels Ltd., Cuttack	40	I
	Aarti Steels Ltd, Athgarh	30	II
	Maheswari Ispat Ltd., Khuntuni	25	II
	Miathan Ispat Ltd, Jhakapur	16	II
Kalahandi	Vedanta Aluminium Ltd, Lanjigarh	75	I

District/Area	Name	Capacity	Category
Koraput	NALCO Ltd, Damanjodi	55.5	I
Balasore	Birla Tyres	07.5	II
Keonjhar		33MW	
	Orissa Sponge Iron Ltd, Palaspanga	12	II
	Sree Metaliks, Loidapara	10	II
	Patnaik Steel and Alloys, Dubuna	07	II
	Deepak Steel and Power Ltd, Topadihi	04	II
Category I		7331 MW	
Category II		276.5 MW	
Total Thermal Power		7607.5 MW	
Category III	Power generated from waste heat is not		
	Included in this paper	378.625 MW	
Proposed thermal power plants: (State Pollution Control Board (SPCB) has cleared them) for establishment:			
Angul Area		8090MW	
Jharsuguda Area		5454.5MW	
Other Areas		4618.5MW	
Projects Pending for clearance with SPCB		7698MW	

Angul-Denkanal area currently account for 61.33% of total thermal power generation

622 million tonnes in 2005 with an annual increase of 38 million tonnes (equivalent to 138 million tones of CO_2) since 1995 (carbon stock was 6, 245 million tones).

It has been estimated that the CO_2 removal by India's forests and tree cover was enough to neutralize 9.31% of India's total emissions (CO_2 equivalent) at 2000 level of emission. In 2000 India's level of GHG emissions was estimated as ~1485,

From these four industrial sources, Orissa emits >164 million tonnes of Green House Gas (GHG), which is not a pollutant as per the provisions of Air (Prevention and Control) Act, 1981. Currently the estimate of India's emissions amount to > 2115 million tonnes (Shukla, 2006).Thus Orissa's emissions account for about 7.8 % of India's emissions. The emissions from other industries (fertlizer, tyre, etc) transport, agriculture and households etc have not been taken into the calculation. Therefore, the GHG emission is huge. In the coming decade, the GHG emission will be doubled gradient, type of forests, the succession stage of the forest and the ground vegetation. However, this shows the importance of forests and tree cover and the urgent need to increase forest and tree cover in India. The grasslands, scrub lands, crops also neutralize GHGs and this has not been estimated separately. The other GHG sinks are soil, water bodies and planktons in the water bodies including Bays and Oceans.

Kishwan et al., (2009) have also estimated component-wise carbon in India's forest biomass and in soil in 1995 and 2005 (Table 5). The total estimate for 2005 was 3755.811 million tonnes of carbon in soil amounting to 55.4723 tonnes per ha. The total estimate amount of carbon in forest biomass was 2865.739 million tonnes amounting to 0.3310 tonne of carbon per ton of biomass and 42.32 tonnes of carbon per ha of biomass.

Net primary productivity (NPP) measurement method

The soil organic carbon pool (SOC) in Orissa amounts to about 5.93% of India's SOC pool and the detail forest type-wise distribution is given in Table 5.The tropical dry deciduous and moist deciduous and littoral and swamp forests contribute significantly to India's SOC pool. This situation can be improved if efforts for forest conservation will be effective million tonnes (Sharma et al., 2006; Shukla, 2006). 9.31% of it amounts to ~138.25 million tonnes. Thus it appears 78.37

Table 3: The estimate of currently GHG emissions from thermal power plants and from other industrial sources in Orissa

Industry	Capacity of Production	Amount of GHG	Future (10 years)	(Additional MW/ million t)
	(MW/Million Tonnes)	Emission (Million tonnes)	Projection	GHGEmission
Thermal Power	7607.5MW	53	25,860MW	180.2
Iron and Steel	17	59	20	70.7
		(MOU Signed)	(60)	(212)
Ferro – Chrome	1	2.1	3	6.3
Aluminium	2	44	3	66
Cement	7	6.3	7	6.3
Total		164.3	+	(329.5)

Table 4: Forest and tree cover of the country in 2007

Class	**Area**	**% of Geographical**
	(million ha) area (G. A.)	
Forest Cover		
VDF (very dense)	8.35	2.54 Note: 55.51
		million ha lies<1000m,
		11.67 m.ha.
MDF (medium dense) 31.90		9.71 <1000 to
		3000m and 1.91 m.ha
		<3000m
OF (open)	28.84	8.77 altitude.
Total Forest Cover	69.09	21.02
Tree Cover	09.28	02.82
(Tree patches<1ha		
With canopy		
density >10%)		
Total Forest and	78.37	23.84
Tree Cover		
Non-forest: Scrub	04.15	01.26
Non-forest	255.49	77.72
Total G.A.	328.73	100

million ha of forest with 12% tropical evergreen category, 34% tropical moist deciduous, 30%tropical dry deciduous, 5% tropical thorn, 6% subtropical pine, and 11% Himalayan temperate category and tree cover could neutralize 138.25 million tonnes of GHG per year amounting on the average to 1 ha of forest and tree cover neutralizing 1.764 tonnes of GHG per year (FSI Report, 2009; Kishwan et al., 2009; Melkenia, 2009). This is a gross estimate and the site specific estimates would be different.

This average value appears to be an underestimate as it does not differentiate the rate of primary productivity on an altitudinal

In field condition in 40-50 year old forest of tropical-subtropical and temperate regions, the annual growth on dry weight basis (NPP) would be about 13-30 tonnes/year/ha in mid latitudes and below 2000m altitude (Lieth, 1972, 1977; Singh and Joshi, 1979; Swain and Behera, 1998; Karmachari and Singh, 1992, Dash and Dash, 2009) and considering 40 % of the NPP as carbon (C) (Kishwan et al., 2009),

Table 5: Soil organic carbon pool estimates (0-30cm) in India's forests (million tonnes in 2005), estimates for Orissa in parenthesis (Based on Kishwan et al., 2009)

Forest Type	Area in 000ha in 2005	Mean Soil Organic Carbon(SOC)	Total SOC in 2005
Littoral and swamp forest	481(234.5)	71.062(71.062)	34181.021(16664.039)(48.75%)
Tropical dry deciduous	19156(2833.59)	34.195(34.195)	655037.332(96894.61)(14.79%)
Tropical moist deciduous	24284(1954.2)	55.009(55.009)	1335848.398(107498.59)(08.04%)
Tropical semi-evergreen	2946(33.221)	54.625(54.625)	160925.000(1814.70)(1.12%)
Plantation/Tropical open	-(53.252)	20.375 (Tropical thorn forest)(20.375)	-(185.009)
Total	67706 including other types(4885.5)		3755811.310 (223056.95)(05.94%)

Table 6: District wise forest cover in Orissa in 2007 and Green House Gas (GHG) mitigation potential. (Area in km^2, (1 km^2=100ha)

District	Geographical Area(GA)	Very dense, Moderately dense/Open forest	Total Forest Cover	Scrub(ha)	GHG Absorption Potential as per FSI Paper(2009) million tonnes per ha)	GHG Absorption as per Net Primary Productivity (million tonnes per ha.)	Forest cover as % of GA
Angul	6,375	364, 1355, 950	2,669	133	0.471	5.07-11.7	41.87
Baleswar	3,806	23, 125, 153	301	56	0.053	0.57-1.32	7.91
Baragarh	5,8 37	176, 373, 348	897	79	0..158	1.70-3.95	15.37
Bhadrak	2,505	01, 10, 13	24	0.0	0.0042	0.046-0.105	0.96
Balangir	6,575	72, 222, 640	934	221	0.165	1.77-4.11	14.21
Boudh	3,098	263, 544, 448	1255	77	0.221	2.38-5.52	40.51
Cuttack	3,932	46, 216, 397	659	124	0.116	1.25-2.90	16.76
Deogarh	2,940	155, 620, 566	1341	15	0.236	2.54-5.90	45.61
Denkanal	4,452	172, 343, 829	1,344	144	0.237	2.55-5.91	30.19
Gajpati	4,325	90, 1523, 872	2,485	259	0.438	4.7-10.93	57.46
Ganjam	8,206	156, 1,038, 771	1965	750	0.347	3.73-8.64	23.95
Jagatsinghpur	1,668	0.0.07, 15	22	01	0.0039	0.042-0.097	1.32
Jajpur	2,899	06, 69, 180	255	53	0.045	0.484-1.121	8.80
Jharsuguda	2,081	04, 143153	300	44	0.0529	0.57-1.32	14.42
Kalahandi	7,920	370, 743, 1,193	2,306	475	0.41	4.38-10.14	29.12

District	Geographical Area(GA)	Very dense, Moderately dense/Open forest	Total Forest Cover	Scrub(ha)	GHG Absorption Potential as per FSI Paper(2009) million tonnes per ha)	GHG Absorption as per Net Primary Productivity (million tonnes per ha.)	Forest cover as % of GA
Kendrapara	2,644	81, 87, 26	194	06	0.0342	0.369-0.853	7.34
Keonjhar	8,303	308, 1,410, 1,511	3,229	51	0.569	6.135-14.20	38.89
Kandhamal	8,021	663, 2,652, 2,169	5,484	367	0.967	10.42-24.12	68.37
Khurdha	2,813	23, 190, 162	375	156	0.066	0.712-1.65	13.33
Koraput	8,807	104, 718, 856	1,678	40	0.296	3.19-7.38	19.05
Malkangiri	5,791	157, 712, 1326	2,195	31	0.387	4.17-9.65	37.90
Mayurbhanj	10,418	1,340, 1,711, 939	3,990	34	0.703	7.58-17.55	38.30
Nawapara	3,852	85, 481, 671	1,237	116	0.218	2.35-5.44	32.1
Nawrangapur	5,291	188, 462, 485	1,135	27	0.200	2.16-4.98	21.41
Nayagarh	3,890	187, 975, 504	1,666	225	0.294	3.165-7.33	42.83
Puri	3,479	0.0, 56, 39	95	46	0.167	0.180-0.418	2.73
Rayagada	7,073	456, 901, 1,769	3,126	476	0.551	5.939-13.75	44.20
Sambalpur	6,657	533, 1,7351, 039	3,307	49	0.583	6.283-14.54	49.68
Sonepur	2,337	03, 200, 121	324	69	0.0571	0.615-1.425	13.86
Sundargarh	9,712	1,047, 1,773, 1,243	4,063	144	0.716	7.719-17.87	41.83
Total	1,55,707	7,073, 21,394, 20308	48,855	4,852	8.62	92.824-214.89	31.38
						Mean=153.85	

the value comes to 5.2 to 12 tonnes of NPP-C per ha. This in terms of CO_2 sequestration (3.68 tonne of carbon dioxide=1 tonne of carbon), amounts to about 19- 44 tonnes of CO_2, which might have been used. This estimation is significantly higher than the estimation made in FSI report. One hectare forest may possess about 1000 trees of about 40-50 year old. 1000 trees hold 19-44 tonnes of CO_2 in NPP (average 19-44/1300= 0.019 -0.044 tonne =19-44 kg CO_2 per tree per year).

Table 6 gives data on forests cover and estimation of GHG neutralization potential of forest cover in different districts of Orissa.

Sink potential from forest source in Orissa

Orissa has 155, 707 km^2 geographical areas which constitutes 4.74% area of the country. The recorded forest area of the state is 58, 136km^2 (37.33% of the geographic area) but the forest cover in the state based on October-December, 2006 satellite data interpretation is 48, 855 km^2 (48, 855, 00ha, (31.38% of the geographic area). In terms of forest canopy density classes, Orissa has 7, 073 km^2 very dense forest, 21, 394 km^2 moderately dense forest and 20, 388 km^2 open forest (FSI, 2009).

Out of 30 districts, only 12 districts of Orissa have forests cover exceeding the National target of 33% forest cover for each district. Another 4 districts have forest cover exceeding 23.8% but less than 33% and 14 districts have forest cover less than 23.8%.Some districts such as Bhadrak, Jagatsinghpur and Puri have forest cover of less than 5% and districts such as Baleswar, Kendrapara and Jajpur have forest cover less than 10% of the geographical area of the district. These are important districts from industrial development point of view. The GHG sink potential is negligible in these districts.

Concentration of Industries

The industries have been concentrated in districts such as Angul, Sundargarh, Jharsuguda, Denkanal, Balasore, and Jagatsinghpur, Jajpur and Sambalpur for the availability of coal, other raw materials and water. The GHG sink potential of Jagatsinghpur, Jajpur, Jharsuguda, Baleswar, Bhadrak, Cuttack, Puri, Kendrapara and Balangiri need urgent attention. Further concentration of thermal power plants in Angul, Denkanal, Jharsuguda are to be avoided. As per the calculation in Table 6, the average forest GHG sink potential of Orissa is 8.62 million tonnes, calculated on the basis of FSI report and 154 million tonnes as per NPP calculation basis. However, there is a huge negative imbalance considering the amount of GHG emissions in Orissa.

Conclusion

Deforestation is one of the largest contributors to climate change. Rain forests provide countless services to humanity, often unnoticed.50% of world's biodiversity are found in rainforests. The Amazon forest released 20 billion tonnes of moisture every day, helping to water crops around the world. Healthy rainforests absorb about 10% of man's carbon emissions every year. Deforestation releases more CO_2 than all the transport. Saving the rainforests will be an essential first step.

About 50% of best-selling products in supermarkets of Developed countries contain palm oil, linked to rainforest clearance in Southeast Asia. Brazilian farmers make $3000 / ha by clearing forest and growing soybeans for export. Subsistence agriculture in Africa clears forests. Rainforests may disappear in next 50 years-How you shop determines the future of rainforests. >40 million dollars may be needed over the next five years to tackle the problem. (Source: The Prince's Rainforests Project: Rain Forests: The Burning Issue, 1996).

Five million hectares of tropical forests are lost every year-larger than the size of England. The situation in India for forest conservation is improving; a very positive situation (FSI, 2009). However the situation in all states is not the same. Constant effort to conserve forests and to create public awareness is required. Planting trees and increasing tree and forest cover is key to increase sink area.

Arresting climate change will require transforming the global economy from a high carbon to low carbon energy base. High carbon base include coal and petroleum. Low carbon base is corn oil, Jatropha oil, Biomass gasification. No carbon base is solar, wind, hydroelectricity, hydrogen and nuclear. For sustainable development and to prevent climate change, switching over to eco-friendly energy source on a large scale is urgently required to address the call of the earth. The alternative energy source should get top priority. There should be rider on consumerism if we want to leave some resources for the future generations and make the earth habitable for them. Environmental discipline at individual/family, state and national level is need of the time.

Regulatory measures

Carbon dioxide, a green house gas is not considered as a pollutant in India and in most of the countries. It should be considered as a pollutant if the emission concentration is very high than its natural range of occurrence in the atmosphere. Environmental Protection Agency (EPA) in USA has declared carbon dioxide as a pollutant and has brought it under environmental regulatory provisions (Raha and Mishra, 2009). They also have cited a legal case, which has been filed at the New South Wales, Australia Land and Environment Court at curbing carbon dioxide pollution from coal fired thermal power plants. If the Australian court declares CO_2 as a pollutant, the case will have a very positive step to control climate change. In India, thermal power from coal based plants is the most important energy source. More than 80 mega thermal power plants operate in India and it is right time to debate whether carbon dioxide should be considered as a pollutant in India.

The mean estimates of GHG neutralization potential of the forests in Orissa varies between 8.62 to 154 million tonnes. Unless effective steps are taken to neutralize the GHG, the situation in future will be dangerous. There is need to implement intensive and extensive plantations, other avenues of GHG emission reduction and neutralization. The state's policy of concentrating of thermal power plants in Angul-Denkanal area and Jharsuguda area require immediate review. The carrying capacity (the assimilatory potential) studies of these areas in particular are immediately required. Fresh water availability is another important aspect. There will be emission of huge amount of other pollutants and huge generation of fly ash in these districts. This requires good planning, governance and implementation of environmental discipline. However the change over to renewable energy source should be planned for a long term basis in accordance with the national policy.

References

Centre for Science and Environment (CSE), New Delhi. 2009. Climate Change: Facts and Politics. p. 125, CSE, New Delhi.

Dash, M. C. 2010. Climate Change: Facts and Politics (In Oriya language), Bigyan Diganta, 6th issue, pp. 3-14, Orissa Science Academy, Bhubaneswar.

Dash, M. C. and Dash, S. P. 2009. Fundamentals of Ecology. Tata-McGraw-Hill Publ., New Delhi.

Dash, M. C. 2007. Save the Planet from Global Warming. Society of Geoscientists and Allied Technologist Bulletin. 8 (2) : 43-59.

Dash, M. C. (Edited). 2006. The State of Environment of Orissa, MOEF, New Delhi and SPCB, Bhubaneswar.

Fatesingh, H. S. 2010. Energy Generation and Conservation, (In Oriya), The Samaj, 26th December, 2010 issue, Cuttack.

Forest Survey of India Report. 2009. Government of India, MoEF, New Delhi.

Jain, M. K. 2010. Suitability of Fly ash as a mine void- fill material-A critical Review, Minenvis. 64: 1-6.

Karmachari, S. B. and Singh, S. K. 1992. Biomass and Net production of Teak (Tectona grandis) in dry tropical region in India. Forest Ecology and Management. 55: 233-247.

Kishwan, Jagdish, Pandey, R. and Dadhwal, V. K. 2009. India's Forest and Tree cover: Contribution as a Carbon Sink, Indian Council of Forestry Research and Education, Dehradun. pp. 1-12.

Lieth, H. 1972. The Net Primary Production of the Earth with special emphasis on the Land Areas, In Perspectives on Primary Productivity of the Earth (edited by Whittaker, R. H. (Ed.), Symposium at the American Institute of Biological Science, Second National Congress, Miami, Florida. USA.

Lieth, H. 1972. Modeling the Primary Productivity of the World, Tropical Ecology. 13: 125-130.

Lieth, H. 1977. The Gross Primary Productivity Pattern on the Land Vegetation: a first attempt, Tropical Ecology, 18, 109-115.

Melkania, N. P. 2009. Carbon Sequestration in Indian Natural and planted forests, Indian Forester. pp. 380-392.

Paribesh Samachar, 2010. Biodiversity-Ecosystems Management and the Green Economy, Volume XV and XVI, pp. 1-25, State Pollution Control Board, Orissa, Bhubaneswar.

Raha, D. and Mishra, P. C. 2009. First ever Legal action on carbon dioxide pollution and its future consequences. The Ecoscan. 3 (3and4) : 209-210.

Rain Forests: The Burning Issue.1996. The Prince's Rainforests Project, Forest Stewardship Council, London. p. 48.

Singh, J. S. and Joshi, M. C. 1979. Primary Production, In Grassland Ecosystems of the World Ed. R.T.Coupland. Cambridge University Press, London. pp. 197-218.

Swain, S. and Behera, N. 1998. Biomass and Net production of Tectona grandis (Teak) in a regenerating teak forest. Ecology, Env. and Conservation. 4 (1and2) : 1-7.

Sharma, S., Bhattacharya, S. and Garg, A. 2006. Greenhouse gas emissions from India: a perspective. Curr. Sci. 90 (3) : 326-333.

Shukla, P. R. 2006. India's GHG emission scenarios: Aligning development and stabilization paths. Curr.Sci. 90 (3) : 384-395.

UNEP, 2007. A Report of the Intergovernmental Panel of Climate Change, 2007. (4 volumes), UNEP, Switzerland.

17

Carbon Sequestration and Role of Earthworms in Indian Land Uses: A Review

Introduction

Carbon sequestration studies have generally involved the vegetation, particularly the forest vegetation and the estimation of soil organic carbon in forest systems (Sharma *et al.,* 2006, Shukla, 2006, Kishwan *et al.,* 2009, Melkania, 2009, Tandon *et al.,* 2013).There have been scientific evidences on the ameliorative property of the vegetation with regard to abatement of pollution (Mishra *et al.,* 2010; Mishra *et al.,* 2011).Vegetation can neutralize CO_2 emission, acting as a potential sink (Ravindranath, 2007; Ravindranath and Ostwald 2008; Ravindranath *et al.,* 2008; Rawat and Kishwan, 2008; Dash, 2010). Recently, in the context of climate change, interest has been generated among scientists to understand the mechanism of C-sequestration among soil biodiversity (Dash, 2013). This review paper discusses carbon sequestration in Indian natural forest systems, rubber plantations in Tripura and coal ash mund reclamation plantations with associated soil biodiversity, especially earthworms.

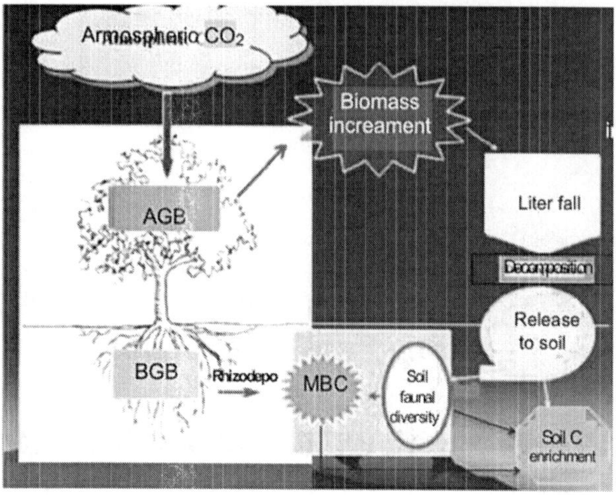

Figure 1: Carbon flow model involving soil biodiversity (Note: AGB-above ground biomass, BGB-below ground biomass, MBC-microbial biomass carbon)

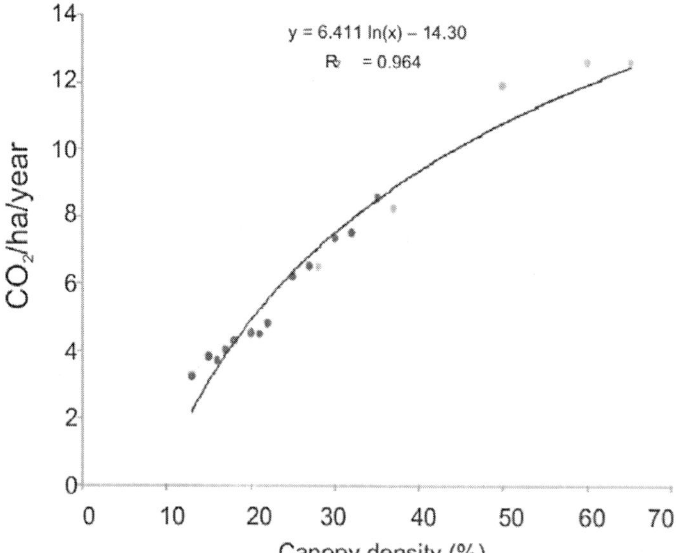

Figure 2: The relationship of forest canopy density and CO_2 absorption (Note: From the regression curve, it appears that canopy density 15-40 % absorbs about 3.5 to 10 tonnes of CO_2 per ha per year and canopy density 50 to 70% absorbs about 12-13 tonnes of CO_2 per ha per year)

Figure 3: An Ash Mound at Hirakud

Figure 4: Plantations in 5 year old site

Fig. 1 gives a conceptual model of carbon sequestration in the terrestrial ecosystem involving soil biodiversity.

In a specified time and habitat, the carbon input and flow should be measured to get an overall picture so that the role of each component in the ecosystem can be assessed and appropriate conservation measures can be taken. Soil biodiversity component has not received much attention with regard to carbon sequestration.

Indian Forest Cover

As per the Forest Survey of India Report (2011) the estimate of forest cover in India as per the satellite imagery methods is given in Table 1. Forest canopy provides CO_2 absorption surface area, protects the soil from rain and sun light directly heating it and after litter fall carbon compound gets incorporated into the soil and provide energy for the activities of soil biodiversity.

Forest and tree cover of India was 78.37 million ha in 2009 (FSI, 2011), which was 23.84% of the geographical area and this includes 2.82% tree cover. MDF should be converted to VDF and OF to MDF, and Scrub to OF to increase canopy cover.

India's level of green house gas emissions was estimated at about 1485 million tonnes in 2000 and forests could neutralize about 9.31% (138.25 million tons) (Sharma *et al.,* 2006, Shukla, 2006)). India's 78.37 million ha forest and tree cover with 12% tropical evergreen category, 34% tropical moist deciduous, 30%tropical dry deciduous,

Table 1: Estimate of Indian Forest Cover in million ha (FSI, 2011)

Class	Area (million ha)	% of Geographical Area (G.A.)
VDF (very dense)	8.35	2.54
MDF (medium dense)	31.90	9.71
OF (open)	28.84	8.77
Total Forest Cover	69.09	21.02
Tree Cover	09.28	02.82
(Tree patches<1 ha with canopy density >10%)		
Total Forest & Tree Cover	78.37	23.84
Non-forest: Scrub	04.15	01.26
Non-forest	255.49	77.72
Total G.A.	328.73	100

Table 2: Region-wise forest cover (sq km) in India (based on FSI Report 2011)

Region	Geog.Area	Total Forest Cover	% of GA	Scrub	(Name of States/UT)
South (5)	6, 36,263	1, 18,284	18.59	14,891	Kerala,Karnataka,Tamilnadu,
					Pondicherry,A.P.
North (9)	7, 72,644	83,126	10.76	4,561	(J&K, Punjab, HP, Haryana, Delhi,
					Ukhand, Chandigarh, UP, Bihar
West (4)	8, 49,676	80,205	09.44	0,272	(Maharashtra, Goa, Gujarat,
					Rajasthan)
N-East (8)	2, 62,179	1, 73,182	66.05	919	Arunachal, Assam, Manipur,
					Meghalaya, Mizoram, Nagaland,
					Sikkim, Tripura)
South East (3)	3, 24,153	82,060	25.32	5,544	(Jharkhand, WB, Odisha)
Central (2)	4, 43,436	1, 31,878	29.74	2,263	(MP, Chhatisgarh)
Islands (4)	8,884	6,8 83	77.48	03 .	(A and N, Dadar and Nagerhavelli,
					Daman and Diu, Lakhshadeep)
Total	3,287,263	6, 77,033	20.60	38,475	
Current for India	(2007-2009)	(21. 02FC+2.82TC=23.84%FC&TC)			

Table 3: Vegetation in 5 year old reclamation site

Tree species	Shrub species		Herb species	
1. Acacia moniliformis	1.	*Abutilon indicum*	1.	*Achyranthes aspera*
2. Delonix regia	2.	*Chromolaena odorata*	2.	*Alternanthera sessilis*
	3.	*Helicteres isora*	3.	*Andrographis paniculata*
	4.	Lantana camara	4.	*Clitoria ternatea*
	5.	*Woodfordia fructicosa*	5.	*Cyperus rotundus*
	6.	*Calotropis procera*	6.	*Eragrostis coarctata*
	7.	*Euphorbia hirta*		
	8.	*Evolvulus alsinoides*		
	9.	*Phylanthus fraternus*		
	10.	*Tephrosia vellosa*		
	9.	*Calicotome spinosa*		

Table 4: Vegetation in 10 year plantation site

Sida cordifolia	*Sida cordifolia*		*Sida cordifolia*	
1. Acacia moniliformis	1.	*Chromolaena odorata*	1.	*Achyranthes aspera*
	2.	*Woodfordia fructicosa*	2.	*Alteronanthes sessilis*
2. Cassia angustifolia	3.	*Ziziphus oenoplia*	3.	*Anmdrographis paniculata*
3. Delonix regia	4.	*Malva sylvestris*	4.	*Andropogon muricanus*
	5.	*Macuna pnuriens*	5.	*Cynodon dactylon*
			6.	*Eragrostis coarctata*
			7.	*Evolvulus alsinoides*
			8.	*Hemidesmus indicus*
			7.	*Euphorbia hirta*
			8.	*Evolvulus alsinoides*
			9.	*Phylanthus fraternus*
			10	*Sida cordifolia*

Table 5: Vegetation of natural site

Tree		Shrub		Herb	
1.	*Adina cordifolia*	1.	*Cinnamomum cassia*	1.	*Achyranthes aspera*
2.	*Aegle marmelos*	2.	*Cinnamomum toenla*	2.	*Aganosma caryophyllata*
3.	*Albizzia labbek*	3.	*Chromolaena odorata*	3.	*Abutilon indicum*
4.	*Anacardium occidentale*	4.	*Gardenia gummifera*	4.	*Andrographis paniculata*
5.	*Artocarpus integrifolia*	5.	*Holarrhena pubescens*	5.	*Hemidesmus indicus*
6.	*Azadirachta indica*	6.	*Lantana camara*	6.	*Tephrosis purpurea*
7.	*Butea monosperma*	7.	*Mava sylvestris*	7.	*Tridax procumbens*
8.	*Citrus medica*	8.	*Woodfordia fructicosa*		
9.	*Garuga pinnata*				
10.	*Mangifera indica*				
11.	*Pongamia pinnata*				

Table 6: Biomass stock and carbon pool of 5 and 10 year old reclamation sites (t/ha) (Based on Tondon *et al.*, 2013). (NF-natural forest)

	Natural	5 year old	10 year old
	Site	site	site
Stock AGB	491	215	246
Stock AGC	196.4	86	98.4
Annual Biomass- C increment			2.33
Annual Litter production			0.47
Release of Litter Carbon (K=0.7)			0.329
Stock- Below Ground Carbon (BGC)			20
Release of carbon in Rhizo-exudates			32kg/ha
Support of microbial biomass			22kg/ha

Table 7: Gives data on carbon sequestration in rubber plantation (2002-2003) in north-east region of India. Carbon sequestration ('000 t) by Rubber Plantations of north-east (Dey, 2005) and coal ash mund at Sambalpur, Odisha (Tandon *et al.*, 2013)

State Plantation		**Area in ha Soli**	**Carbon Store from**	**Annual Caddition (AGB & BG) under-growth (T per ha) (T per Ha)**
Tripura	28,853	Rubber	66.40	2.4
Based on Dey, 2005				
Assam	13,202	Rubber	69.59	2.4
Meghalaya	4,586	Rubber	98.386	2.39
Nagaland	2087	Rubber	108.0	2.39
Mizoram	696	Rubber	89.65	2.44
Manipur	1708	Rubber	59.95	2.4
Arunachal Pradesh	372	Rubber	156.98	23.92
Central Himalayas				31.276 (Based on Bhadauria *et al.* (2012)
Sambalpur, Odisha	41.7	Ashmund	0.0 279	2.79kg (Based on Tondon *et al.*, 2013)
		Reclamation (10 year old)	(27.9kg)	

05% tropical thorn, 06% subtropical pine and 11% Himalayan temperate category taking all altitudes together and all type of canopy cover; very dense, medium dense and open taken together could neutralize 138.25 million tonnes of GHG per year amounting on the average of 1 ha of forest and tree cover neutralizing 1.8 tonnes of GHG per year (estimation based on satellite imagery mapping method, FSI Report, 2009, Kishwan *et al.,* 2009, Melkania, 2009-). Table 1 gives region-wise forest cover in India (estimated as per FSI report 2009).This was a gross estimate and the site specific estimates would be much different.

Site specific estimations by primary productivity method are significantly higher than the estimation made using FSI report. In Table 2 region-wise forest cover is given.

The forest cover in north, south and western part of the country is far away from the national mandate of 33% of land area under forest cover. Special efforts are required to achieve the goal.

The carbon stocks stored in country's forests and tree cover and soil have been estimated as 6, 245 and 6, 622 million tonnes in 1995 and 2005 respectively with an annual increase of 38 million tonnes (equivalent to 138 million tonnes of CO_2). (Kishwan *et al.,* 2009, Melkania, 2009).

Estimates by the Net Primary Productivity method (NPP method) in different sites indicate that the estimate made as per FSI report is an underestimate (Dash, 2009, Behera and Mishra, 2012).

Net Primary Productivity (NPP) Method: In field condition in a 50-70 year old forest of tropical-subtropical and temperate regions, the annual growth on dry weight basis (NPP) would be about 13-30 tonnes/year/ha in mid latitudes and below 2000m altitude (Lieth, 1972, 1977, Singh and Joshi, 1979, Swain and Behera, 1998, Karmachari and Singh, 1992, Dash and Dash, 2009, Behera and Mishra, 2012) and considering 34 to 45 % of the NPP as carbon (C) (Kishwan *et al.,* 2009, Tandon *et al.,* 2013)), the value comes to 4.42-5.85 to 10.2-13.5 tonnes of NPP-C per ha/yr. This in terms of CO_2 sequestration (3.67 ton of carbon dioxide=1 tonnes of carbon), amounts to about 16-21 to37-49.5 tonnes of CO_2.

Fig. 2 shows the relationship of carbon absorption by the natural forests and the canopy density. The work is based on few natural forest stands near Hyderabad (Behera, 2010 - unpublished data).

Plantation Forests

Plantation on coal-ash mound

About three thousand tonnes of coal ash amended with few tonnes of field soil per day is dumped in ash mound covering an area of 41.7 ha. Plantations have been taken up over the ash mound to stabilize the system and to prevent erosion and leading to the reclamation. Fig. 2-5 shows the ash mund, 5 year and 10 year old reclamation sites and a natural site near to the ash mund area.

In a study of coal ash mund reclamation study at Hindalco industries, Hirakud, Odisha with *Acacia moniliformis, Cassia angustifolia and Delonix regia* plant species, Tandon *et al.* (2013) found that in 10 year old plantation the C-sequestration per year amounted to 2.79tC (10.23 t CO_2) per ha. Organic carbon pool analysis of ash mound soil under plantation revealed a carbon enrichment of 27.9 kg/ha/yr due to steady flow of carbon from litter decomposition.

Table 3 gives the data on carbon sequestration in ash mund (Tandon *et al.,* 2013). On the basis of measuring diameter of trees at breast height and using the allometric equation developed by Brown *et al.* (1989), the above ground biomass stock was estimated and the carbon content in the biomass was also estimated. Soil-ash samples were analysed for presence of earthworms and their biomass was measured and carbon content in the biomass was estimated.

The constituents of fly ash are oxides of iron, silica, potassium, alumina, traces of sulphur, phosphorus, cadmium, chromium, lead, mercury, arsenic, zinc, boron, selenium, strontium etc. Indian ash contains calcium, sodium, magnesium and titanium along with the above said elements. Carbon and nitrogen are usually present in negligible quantities. However, available phosphorus and potassium levels are generally higher as compared to the parent material *i.e.,* coal. Fly ash being an amorphous ferro-alumino-silicate mineral contains all the naturally occurring elements which occur in coal. The major elements in fly ash are Si, Al and Fe.

Generally fly ash contains higher concentration of essential plant nutrients like sodium, potassium, calcium, magnesium, boron, sulphates and other nutrients except nitrogen in comparison to soil. Fly ash munds can support vegetation with soil amendments and with water availability. The study was carried in 5 year and 10 year old plantation

Figure 5: Plantation in 10 year old site

Figure 6: Natural site near to ash mund reclamation site

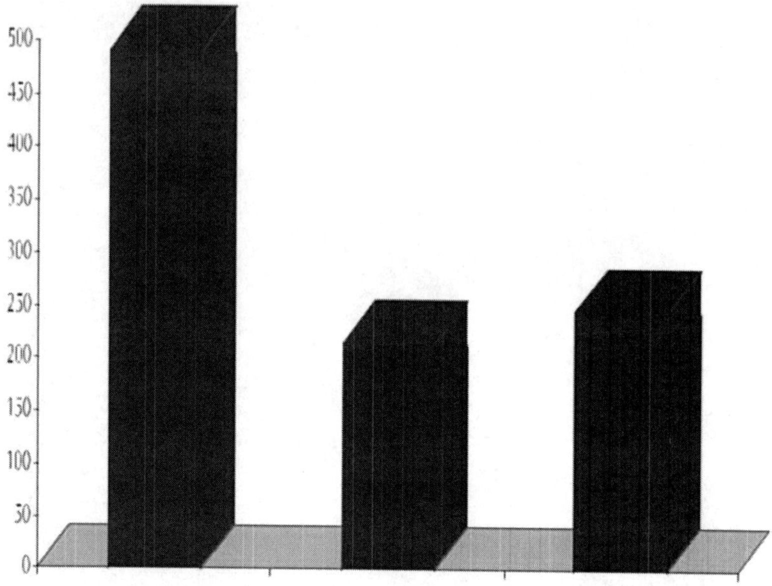

Figure 7: Annual increment of biomass in Above Ground Biomass (ABG) _ biomass stock

Biomass calculation: Y = 38.49018 - 11.7883D+1.926D2, (Brown *et al.* 1989) where, Y= Biomass (kg), D= DBH (cm), Biomass annual increment in plantation on ash mound: (10 yr biomass - 5 yr biomass) /5 =6.2 t/ha/yr

Table 8: Comparison of herbaceous species in the three sites

Parameter	Natural forest	Plantation site	
		5 year	**10 year**
Species richness	7	12	10
Species diversity (H)	1.84	2.14	2.06
Species dominance (cd)	0.153	0.131	0.149
Species density (no/m2)	13	25	25

(Based on Tondon *et al.,* 2013). Fig. 7 gives annual increment of biomass in Above Ground Biomass (ABG) stock.

sites and the data were compared with natural forests situated close to ash mund sites. The plant species were identified using Brahmam. and Saxena (1995)

The biomass calculation of the Above Ground Biomass (AGB) stock of natural site (NF), 5 year and 10 year old site are given in Fig. 6. The AGB of natural forest (NF) (491 t with 275 t C) is about two times higher than 10 year old plantation (246 t with 162 t C) site and annual increment of biomass was 6.2 t/ha. The Below Ground Biomass (BGB) in 10 year old plantation and carbon content was respectively 50 t/ha and 20 t/ha. Table 5 gives the vegetation of the natural forest site. Table 6 gives a comparison of three sites. Table 7 gives Biomass Stock and Carbon pool of 5 and 10 year old reclamation sites (T/ha) absorption (Note: From the regression curve, it appears that canopy density 15-40 % absorbs about 3.5 to 10 tonnes of CO_2 per ha per year and canopy density 50 to 70% absorbs about 12-13 tonnes of CO_2 per ha per year)

Rubber plantation in North-East

Carbon sequestration is high in rubber plantation amounting to average 136.15 t C per ha (61.1 t C per ha AGB, and 75.06 t C per ha BGB), equivalent to about 500 t CO_2 per ha. Coal Ashmund with soil amendment is suitable for reclamation with vegetation .Organic carbon built up in soil is slow but expected to increase after few years.

Soil Carbon and Earthworms

Kishwan *et al.* (2009) have also estimated component-wise carbon in India's forest biomass and in soil together in 1995 and 2005. The total estimate for 2005 was 3755.81 million tonnes of carbon in soil amounting to about 55.47 tonnes per ha. The total estimate amount of carbon in forest biomass (AGB) was 2865.74 million tonnes amounting to 42.32 tonnes of carbon per ha and 0.34 ton of carbon per ton of plant biomass. The tropical dry deciduous and moist deciduous and littoral swamp forests contribute significantly to India's SOC pool. This situation can be improved if efforts for forest conservation will be effective. Bhadauria *et al.* (2012) have studied earthworm biomass structure in seven land uses in Bhiri-Banewara village land scape in Central Himalayas and found on the average 11.17g of soil organic carbon (soc) per kg of soil, which amounts to about 31.276 tonne of soc per ha (estimation based on Brady, 2000). In central Himalayas

Table 9: Soil Organic Carbon (SOC) pool Estimates (0-30cm) in India's Forests (million tonnes in 2005), Estimates for Odisha in parenthesis (estimate based on Kishwan *et al.*, 2009).

Forest Type	Area in '000ha 2005	Mean SOC	Total SOC in 2005
Littoral	481	71.062	34181.021
And swamp	(234.5)	(71.062)	(16664.039)
	(48.75%)		
Tropical	19156	34.195	655037.332
Dry deciduous	(2833.59)	(34.195)	(96894.61)
		(14.79%)	
Tropical moist	24284	55.009	1335848.398
Deciduous	(1954.2)	(55.009)	(107498.59)
		(08.04%)	
Tropical	2946	54.625	160925.000
Semi-evergreen	(33.221)	(54.625)	(1814.70)
		(1.12%)	
Plantation /	-		
Tropical open	(53.252)	20.375 (Tropical thorn forest)	-
	(20.375)		(185.009)
Total 67.706 mills. ha including	Other types		3755811.310
(78.37 m.ha Present)		(Odisha- 4.8855 mha)	
			(223056.95)

Table 10: Soil organic carbon pool estimates in 14 natural forests (calculation based on Kishwan *et al.*, 2009)

Sl. No	Forest Type (Group)	Total SOC (Million Tonnes)	
		1995	**2005**
1.	Himalayan dry temperate forest	1122.144	1158.343(.03%)
2.	Himalayan moisttemperate forest	159616.937	175149.1689(4.66%)
3.	Littoral and swamp forest	27216.904	34181.021(0.91%)
4.	Montane wet temperate forest	298233.293	299387.893(7.97%)
5.	Sub alpine and alpine forest	149698.375	153105.661(4.076%)
6.	Sub tropical broad leaved hill forest	22518.8.33	26243.102(0.7%)
7.	Sub tropical dry every green forest	79836.780	81468.766(2.17%)
8.	Sub tropical pine forest	229031.601	238432.15(6.34%)
9.	Tropical dry deciduous forest	623475.447	655037.332(17.44%)
10.	Tropical dry evergreen forest	7021.363	8645.709(0.23%)
11.	Tropical moist deciduous forest	1270222.177	1335848.(35.57%)
12.	Tropical semi evergreen forest	140549.904	160925.000(4.28%)
13.	Tropical thom forest	32681.741	37225.399(0.99%)
14.	Tropical wet evergreen forest	511078.124	549003.366(14.62%)
	Total	35,52,303.628	37,55,811.310(100%)

soc content in soil is about 56% of the average value calculated from Kishwan *et al.* (2009) data (55.47 tonnes)

Soil carbon is largest terrestrial pool of carbon. Dead plant and animal body gets incorporated into the soil and provide parent material of soil organic matter. For estimation of carbon content in animal biomass in soil and plant biomass, 58% and 40% have been taken respectively (Kishwan *et al.*, 2009). Soil biodiversity, especially earthworms ingest these materials, digest part of it and bring fragmentation to the rest- enhance microbial action. Role of soil biodiversity, especially earthworms in the carbon dynamics in the ecosystem is considered important (Dash and Patra, 1972, Senapati and Dash, 1981) but yet to be properly assessed.

Table 11: Site wise earthworm biomass & C-content (kg/ha) habitat biomass kg/ha carbon content kg/ha wet weight dry weight

Habitat	Biomass kg/ha	Carbon content kg/ha	
		Wet weight	Dry weight
Easternghat Region (6.96—81.2)			
1. Pasture, Berhampur, Odisha	60-610	12—122	6.96—70.76
2. Upland pasture, Sambalpur, Odisha	120-700	24—140	13.92—81.2
3. Deciduous Forest, Sambalpur, Odisha	70—285	14—57	8.12—33.06
4. Deciduous Forest, Andhra Pradesh (Estimated)	157.5	31.5	18.27
5. Coal ash-mund plantation (10 year old)	~1 g/ha	<0.5g/ha	<0.3 g/ha
Westernghats, India (32.38—68.14)			
6. Forest	279.2	55.84	32.38
7. Acacia plantation	96.7	56.08	32.52
8. Grassland, Karnataka	587.4	117.48	68.14
Central Himalaya (1.16—78.4) and Tripura			
9. Mixed Forest	410	82	47.56
10. Pine Forest (6year old)	10—400	2—80	1.16-16
11. Village landscape sites (Bhiri - Banswara)-peak	110-2260	88-1808	51-1446
12. Rubber plantation, Tripura	433.7	86.74	50.31
13. Mixed Fruit Plantation, Tripura	560	112	64.9
14. Pine-apple plantation, Tripura	360	72	41.76
Foreign Sites (3.82—69.02)			
15. Rubber Plantation, Cote d'Ivoire)	595	119	69.02
16. Mexico, Rain Forest	342—424	68.4-84.8	39.67-49.18
17. Malaysia, Rain Forest	33—227	6.6—45.4	3.82—26.33
18. Savanna, Ivory Coast (20 years)	490	98	56.84

On the basis of forest cover area, the soil organic carbon pool (SOC) in Odisha amounts to about 6% of India's SOC pool and the detail forest type- wise distribution is given in Table 5. The tropical dry deciduous and moist deciduous and littoral and swamp forests contribute significantly to India's SOC pool. This situation can be improved if efforts for forest conservation will be effective. Table 9

Table 12: Estimate of soil biodiversity carbon in India (mean=2 ton C per ha). Forest type (group) total soil biodiversity C per ha (tonnes)

Forest type (group)	Total soil biodiversity C per ha (tonnes)	
2005	area in 000 ha	C
1. Himalayan dry temperate forest	32	64,000
2. Himalayan moist temperate forest	2447	48, 94,000
3. Littoral and swamp forest	481	9, 62,000
4. Montane wet temperate forest	2593	51, 86,000
5. Sub alpine and alpine forest	2067	41, 34,000
6. Sub tropical broad leaved hill forest	303	6, 0, 6000
7. Sub tropical dry evergreen forest	1248	24, 96,000
8. Sub tropical pine forest	4743	94, 86, 000
9. Tropical dry deciduous forest	19156	3, 83.12,000
10. Tropical dry evergreen forest	165	3, 30,000
11. Tropical moist deciduous forest	24284	4, 85, 68,000
12. Tropical semi evergreen forest	2946	58, 92,000
13. Tropical thorn forest	1827	36, 54,000
14. Tropical wet evergreen forest	5414	1, 08, 28,000
Total	67,706	13, 54, 12,000
		135.412 million tons

and 10 give data on soil organic carbon in Indian forests (estimation based on Kishwan *et al.,* 2009).

Earthworm biomass and carbon content is given in Table 11 and Table 12.

The carbon dynamics is linked to the activity of soil biodiversity functioning, which is very important in many Indian land uses. There is need of popularizing the concept that forests and soil biodiversity are important C-sinks and they are to be protected, soil erosion to be prevented to allow the soil biodiversity to build up their population and afforestation is to be practiced.

Acknowledgement

Professors: P. C. Mishra of Sambalpur University and currently expert member of Green Tribunal, New Delhi allowed us to use his coauthored published data and photographs. We acknowledge his help.

References

Behera, N. 2010. Unpublished data.

Bhadauria, T., Kumar, P., Kumar, R., Maikhuri, R. K., Rao, K. S. and Saxena, K. G. 2012. Earthworm populations in a traditional village Landscape in Central Himalaya, India. *Applied Soil Ecology*. 53: 83-93.

Brady, N. C. 2000. The Nature and Properties of Soil, Prentice-Hall Publishers, New Delhi, 621 pages Brahmam, H. and Saxena, H. O. 1995. The Flora of Orissa, Vol. I to V, Publ. RRL, OFDC, Bhubaneswar. under various Plantation Stages. Final Project Report submitted to Hindalco Industries Limited, Hirakud, 30th April, p. 158. 8 (3) : 19-28.

Brown, S. G., Andrew, J. R. and Lugo, A. E. 1989. Biomass estimation method for tropical forest with application to forest inventory data. *J. Forest Science.* 35 (4) : 881-902.

Dash, M. C. and Patra, U. C. 1972. Density, biomass and energy budget of a tropical earthworm population from Orissa, India. *Review of Ecology and Biology of Soil.* 14: 461-471.

Dash, M. C. 2007. Save the Planet from Global Warming. Society of Geoscientists and Allied Technologist Bulletin. 8 (2) : 43-59.

Dash, M. C. and Dash, S. P. 2008. Conservation and sustainable management of belowground biodiversity: A review on the functional role of soil fauna in Indian ecosystems with particular reference to earthworms. *International J. Ecology and Environmental Science.* 34 (3) : 223-243.

Dash, M. C., Saxena, K. G. and Giri, S. 2009. Vermitechnology for Wasteland Reclamation, Plant Productivity and Composting: A Review in Indian Context. *International J. Ecology and Environmental Sciences.* 35 (2-3) : 163-185.

Dash, M. C. and Dash, S. P. 2009. Fundamentals of Ecology. *Tata-McGraw-Hill Publ.*, New Delhi

Dash, M. C. 2010. A comparison of Industrial Green House gas emissions (GHG) and carbon sink potential of forest vegetation in Orissa, in the context of climate change. *The Ecoscan.* 4 (2 and 3) : 147-152.

Dash, M. C. 2011. Energy demand and carbon budgeting for India.

The Ecoscan: Special issue. 1: 01-05.

Dash, M.C. and Saxena, K. G. 2012. Earthworms in the Himalayan and Western Ghats Region of India, The Bioscan, 7 (1) :1-12.

Dash, M. C. 2013. Sustainable Development: The Journey from Rio-1992 to Rio+20 (2012) and Beyond, International *J. Ecology and Environmental Science.* 39 (1) :1-21.

Dey, A. 2005 (Quoted by Dey, Animesh and Chaudhuri, P. S. 2012. Earthworm Community structure in the pineapple (Ananus comosus) monoculture plantation and pineapple mixed culture plantations of West Tripura, India and Tropical Ecology (In Press). Forest Survey of India, Report, 2011. Government of India, Ministry of Environment and Forests. New Delhi

Karmachari, S. B. and Singh, S. K. 1992. Biomass and Net production of Teak (Tectona grandis) in dry tropical region in India. *Forest Ecology and Management.* 55: 233-247.

Kishwan, J., Pandey, R. and Dadhwal, V. K. 2009. India's Forest and Tree cover: Contribution as a Carbon Sink, Indian Council of Forestry Research and Education, Dehradun. pp. 1-12.

Lieth, H. 1972. The Net Primary Production of the Earth with special emphasis on the Land Areas, In Perspectives on Primary Productivity of the Earth (edited by Whittaker, R. H. (Ed.), Symposium at the American Institute of Biological Science, Second National Congress, Miami, Florida. USA.

Lieth, H. 1977. The Gross Primary Productivity Pattern on the Land Vegetation: a first attempt, Tropical Ecology. 18: 109-115.

Melkania, N. P. 2009. Carbon Sequestration in Indian Natural and planted forests, Indian Forester. pp. 380-392.

Mishra, P. C., Behera, N., Mishra, S. P. and Meher, K. 2011. Ecological Enumeration of Tree Vegetation and Their Contribution in Removal of Atmospheric Pollution Load: A Case Study in an Industrial Complex of Western Orissa, India. *Asian J. Water, Environment and Pollution.*

18

Load of Pollutants and Green House Gas (GHG) in the Atmosphere in India and the GHG Sink Factor

Introduction

Indian Satellite pictures show thick haze and black carbon smoke above Indian atmosphere and other Asian countries. This thick haze and smoke is particularly severe along the Ganges Basin. Biomass burning in rural parts and urban slums of India is the major sources of particulate matter and aerosols, and it is generally accepted that the black carbon smoke over the Indian troposphere is from biomass burning.

The air pollution in large cities such as Kolkata, Delhi, and Mumbai etc is due to transport emissions, biomass and coal burning and other sources. Further, it is estimated that kitchen smoke leads to around 1.5 million premature deaths every year, more than the number of deaths from malaria each year. Millions of Indians use fuel wood, agricultural waste and biomass (cow dung) cakes for cooking and general heating needs. These traditional fuels are burnt in chulha in many parts of India. (Atmanand et al. (2009) and Ganguly et al (2001). These traditional fuel burning releases high levels of smoke, PM**10** particulate matter, NO**X**, SO**X**, polyaromatic hydrocarbons (PAHs), formaldehyde, carbon monoxide and other pollutants (UNEP, 2002,, WHO and UNEP. 2011, *The Times of India*. 3 December 2009). About four lakh people in India die of indoor air pollution and carbon monoxide poisoning every year because of biomass burning in chulha (WHO). Firewood and other biomass burning will not stop, unless electricity or clean burning fuel becomes adequately available and widely adopted in villages and urban centres in India.

Thermal power plants convert energy rich fossil fuels such as coal, natural gas, petroleum products, and biomass such as agricultural waste, domestic trash/waste, etc. into electricity. Coal and lignite accounted for about 60% of India's total installed capacity. India's electricity sector consumes about 72% of the coal produced in the country. The high ash content in coal affects the thermal power plant's emissions and huge amount of ash generation in India. The Indian Ministry of Environment & Forests has mandated the use of beneficiated coals whose ash content has been reduced from about 45% to 34% (or lower) in power plants in urban, ecologically sensitive areas. Coal benefaction industry has rapidly grown in India; with current capacity exceeding 90 million tons. Coal beneficiation creates huge water pollution unless scientifically managed.

A study of 2007 finds that untreated sewage discharge to water bodies is single most important cause for pollution of surface and ground water in India. There is a large gap between generation and treatment of domestic wastewater and maintenance in India. For example none of the municipalities and notified area councils including the capital city Bhubaneswar in Odisha has sewerage treatment plants. Reliable electricity supply to operate the sewerage plants is the basic requirement. The wastewater generated in these areas normally percolates in the soil or evaporates and wastes accumulate in the urban areas causing unhygienic conditions, release heavy metals and pollutants that leach to surface and groundwater. Almost all surface water bodies namely rivers, lakes and ponds in India are severely polluted, and not fit for human use. Water pollution also adversely impacts wetland, rivers and ocean biodiversity. Supply of electricity is essential for addressing water pollution and associated environmental issues in India.

Pollution Load and Green House Gas Load in the atmosphere due to Thermal Power

The basis of estimations of Green House Gas (CO_2) emissions from thermal energy sector and important industrial sectors, which utilize thermal power in the country, is estimated and given in Table 1.

One estimate (Paribesh Samachar, 2010) shows that the pollution load from 1 MW Indian coal based thermal power generation amounts to:

Table 1. GHG Load estimation for India

~19 tons of CO_2/day/ MW
>136 kg of SO2/ day/MW
>7 tons of fly ash/ day/MW
>60 Kg of Particulate matter/ day/ MW

About 300 million tons of fly ash is generated annually in India. (Odisha State ~26-30 million tonnes).

(i) Energy sector: Thermal Power:

1MW generation requires about 20 tons of Indian Coal, which contains on the average 33% C. Thus 1MW generates about 0.79t CO_2 per day per MW. Hence 1000MW generates about 9.02 million tons CO_2/year. India generates ~ 1, 64, 407 MW of electricity from thermal power plants (2014 August) annually & GHG (CO_2) generation is about 1483 million tons.

(ii) Industrial Process sector

(a) Iron & steel production: 1 ton production generates about 3 tons of CO_2 from coal & coke. India produced 82 million tonnes of iron & steel and released about 243 million Tonnes of GHG

(b) Alumina & Aluminum: 1 ton alumina production and 1 ton alumina smelting generate 21.66 ton CO_2. India has the capacity of ~ 8 million tonne aluminum production per year. ~174 million tonnes of GHG

(c) Cement: 1 ton production generates about 0.9 ton CO_2.India has ~ 300 million tonnes annual production capacity at present. ~ 270 million tons of GHG.

Electricity sector in India had an installed capacity of 2, 64, 000 MW by the end of August, 2014.With 4.8% global share in electricity generation, India became the world's third largest producer of electricity in the year 2013. Captive power plants have an additional 3937.5 MW capacity. The total annual generation of electricity from all types of sources was >1102.9 (TeraWatt-hours-_(TWh)_in 2013 (Statistical Review of world energy 2014). Renewable Power Plants (wind, solar, biomass gasification etc) constitute 12.45% of total installed Capacity only (Figure 1).

The per capita average annual domestic electricity consumption in urban areas (about 700kWh) is about 3 times higher than the rural areas. The worldwide per capita annual average consumption of electricity is about 2, 600 kWh and 6, 200 kWh in the European Union. (World Energy Outlook 2011: Energy for All", International Energy Agency. October 2011).

Sources of Electricity in India (Figure 1)

As per the data in the diagram, 59% of the total energy in India comes from coal, 17% from major hydropower, natural gas 9%, nuclear power 2%, 12 % from the renewable source. The breakup of 12% Renewable Energy Source is: small hydro (3804 MW), wind power (21136 MW), solar power (2132 MW), biomass power/co-generation (4014 MW), waste-to-power (107 MW). It is clear that coal is the main source of energy in India.

The planned additional thermal power generation capacity excluding renewable power during the remaining period of the 12th plan period (up to March, 2017) is 90, 925 MW (Central Electricity Authority, Ministry of Power, Government of India. June 2014). About 300 million people in India and about 1.4 billion people in the world do not have access to the electrical energy in their houses. The implications arc enormous. Children cannot study after dark, health clinics do not function properly, and women are burdened with time consuming works such as water collection, pounding grain, leaving them with less time to engage in income generating activities. After gaining access to energy, households generate more income, and become more productive. Manufacturing and other industrial activities and modern agriculture require lot of energy. In view of this energy generation has to increase. Figure 2 gives data on electricity production in India up to 2012.

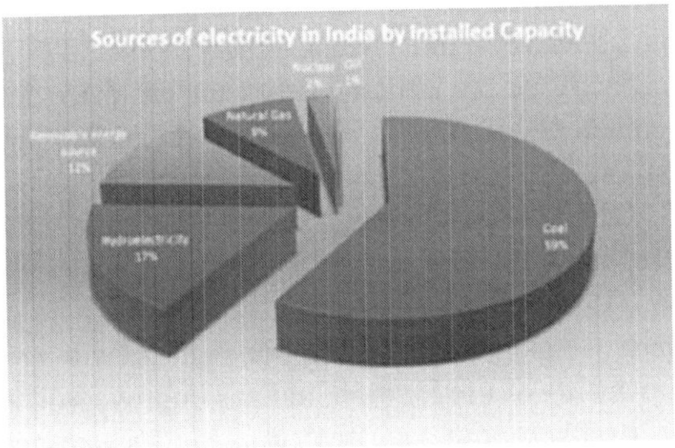

Figure 1. Energy sources in India (from (Google Search 2014 October-Wikipedia, the free encyclopedia)

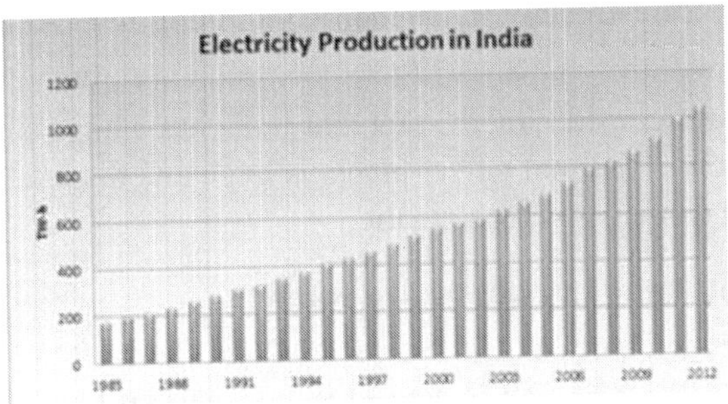

Figure 2. Electricity production in India (From Wikipedia, the free encyclopedia)

The International Energy Agency estimates India will add between 600 Giga Watts (GW) to 1, 200 GW of additional new power generation capacity before 2050-(Uwe Remme et al. (February 2011) International Energy Agency France; OECD).The technologies and fuel sources may make significant impact to India's resource usage and environmental issues (Centre for Energy-Environment Resources Development, Thailand. January 2009). In 2010–11, India's industrial demand accounted for 35% of electrical power requirement, domestic household use accounted for 28%, agriculture 21%, commercial 9%, and public lighting and other miscellaneous applications accounted for the rest. The electrical energy demand for 2016–17 is expected to be at least 1, 392 Tera Watt Hours, with a peak electric demand of 218 GW.The electrical energy demand for 2021–22 is expected to be at least 1, 915 Tera Watt Hours, with a peak electric demand of 298 GW. Considering the fact that the average transmission and distribution losses remain about 32%, India needs to add about 135 GW of power generation capacity, before 2017, to satisfy the projected demand after losses. (Yoginder Alagh, (2011), Technology of India "Transmission and Distribution of Electricity in India Regulation, Investment and Efficiency". OECD. Many experts claim that India's demand for electricity may cross 300 GW, earlier than most estimates. The reasons they give are, (i) India's manufacturing sector is likely to grow faster than in the past, (ii) Domestic consumption demand will increase more rapidly as the quality of life for more Indians improve, (iii) About 125, 000 villages are likely to get connected to India's electricity grid. A demand of 300 GW will require about 400 GW of installed capacity, McKinsey notes. The extra capacity is necessary to account for plant availability, infrastructure maintenance, spinning reserve and losses.

Government of India and State governments give emphasis on thermal power considering the fact that India has a huge coal reserve. The pollution of thermal power plants is huge and disposal of fly ash and bottom ash is very problematic.

Emission of Pollutants and use of water:

1. Suspended Particulate Matter (SPM) - 60kg/day/MW—21.9ton/year/MW (1, 64, 000 MW x21.9 = 35, 91, 600 Ton /Year
2. SO2- >136kg/day/MW—49.64 Ton/yr/MW X (1, 64, 000MW x49.64 =8, 1, 40, 960 Ton/ year)
3. 7 Ton/day/MW—2555Ton/year/MW X 1, 64, 000 MW =419 million Ton/year
4. Water Resource Use- 2.8 Litre/kWh- 24.528 million litres per year per MW,

   For 1, 64, 000MW, water consumption amounts to 40, 22, 592 million litres per year
5. Carbon Dioxide Sink in India : Forest vegetation is the main sink of CO_2.The other sinks are soil, water bodies.

In field condition in 40-50 year old forest of tropical-subtropical and temperate regions, the annual growth on dry weight basis (NPP) would be about 13-30 tonnes/year/ha in mid latitudes and below 2000m altitude (Lieth, 1972, 1977, Singh &Joshi, 1979, Swain and Behera, 1998, Karmachari and Singh, 1992, Dash and Dash, 2009) and considering 40 % of the NPP as carbon (C) (Kishwan et al, 2009), the value comes to 5.2 to 12 tonnes of NPP-C per ha.. This in terms of CO_2 sequestration (3.68 tonne of carbon dioxide=1 toner of One hectare forest may possess about 1000 trees of about 40-50 year old.1000 trees hold 19-44 tonnes of C02 in NPP (average 19-44 kg C02 per tree per year).

Forest and tree cover of the country in 2009

Class	Area (million ha)	% of Geographical Area (G.A.)	
Forest Cover			
VDF (very dense)	8.35	2.54	Note: 55.51 million ha lies <1000m,11.67m.ha.
MDF (medium dense)	31.90	9.71	<1000 to 3000m and 1.91 m.ha <3000m
OF (open)	28.84	8.77	altitude.
Total Forest Cover	69.09	21.02	
Tree Cover (Tree patches<1ha With canopy density >10%)	09.28	02.82	
Total Forest & Tree Cover	78.37	23.84	
Non-forest: Scrub	04.15	01.26	
Non-forest	255.49	77.72	
Total G.A.	328.73	100	

The gradual switch over from coal based energy generation to renewable source in India may help find a sustainable alternative to traditional fuel burning. This should promote conservation of forests and use of manures in crop fields. Government of India's policy is also based on these facts.

India's forest vegetation sink absorbs about 10% of the total CO_2 emission. (Shukla, 2006).The mandate is 33% of India' land cover should have forests. But about 23% of the land covers forest cover of all types (very dense, moderately dense, and open). Algae can absorb CO_2 many times higher than land plants. Using green technology, algal culture can be made in coastal waters, abandoned land, and in man-made industrial (thermal power plants) ponds. Large scale multi species plantation is required. The districts with less than 10% forerst cover require urgent attention.

References

http://www.cea.nic.in/reports/monthly/inst_capacity/aug14.pdf ("Statistical Review of world energy". Retrieved 17 June 2014).

"Monthly All India Installed Generation Capacity Report".

"Growth of Electricity Sector in India from 1947-2013". Central Electricity Authority, Ministry of Power, Government of India. July 2013.

World Energy Outlook 2011: Energy for All". International Energy Agency. October 2011.

Atmanand et al. (2009). "Energy and Sustainable Development-An Indian Perspective". World Academy of Science.

Ganguly et al (2001) "Indoor Air Pollution in India – A Major Environmental and Public Health Concern". Indian Council of Medical Research, New Delhi.

"The Asian Brown Cloud: Climate and Other Environmental Impacts". United Nations Environmental Programme. 2002.

"Indoor air pollution and household energy". WHO and UNEP. 2011.

"Green stoves to replace chullahs". *The Times of India*. 3 December 2009.

"Status of Sewage Treatment in India". Central Pollution Control Board, Ministry of Environment & Forests, Govt of India. 2005.

"Evaluation of Operation And Maintenance Of Sewage Treatment Plants in India-2007". Central Pollution Control Board, Ministry of Environment & Forests. 2008.

"Statistical Review of world energy 2014 workbook".

"Power Generation from Various Renewable Energy Sources". Ministry of New and Renewable Energy. 9 December 2013. Retrieved 10 January 2014.

"New & Renewable Energy Cumulative deployment of various Renewable Energy Systems as on 31/12/2012". Ministry of New and Renewable Energy, Government of India. December 2012.

"Solar". Ministry of New and Renewable Energy, Govt. of India. 2014.

UNEP wins Energy Globe award. Renewable-energy-world.com. 2012

"Facts & Figures, India". Wind Power India. 2011.

19

Environmental Pollution Control Laws applicable to Mining: An Overview

Introduction

Mining and mineral based industries are vital to economic growth of all nations in the world. Mining industry in India has grown manifold over the years and mining operation is usually of large scale and its impact persists over a long period of time.

Mining processes such as drilling, blasting and ore extraction causes air and noise pollution. The problem is further compounded when the ores are crushed and screened. Transportation and handling of mineral also add further to the air and noise pollution. Water pollution due to mine drainage water, ore beneficiation process and tailing disposal are also essential components of mining activity, which has significant potential for water pollution. Mining causes huge land degradation and all kinds of pollution; such as air, water, soil and noise pollution. Prior to mineral extraction, it involves removal of overburden (OB), its transport to the designated OB dump area and its subsequent reclamation. This activity involves air and noise pollution. Leaching of heavy metals from metal mine OB and siltation of surrounding area due to rainfall is also a serious matter of environmental concern as heavy metals can build up in the food chain. The pollution controlling mechanism is money intensive as it requires input of modern technology.

In India the existing laws dealt directly or indirectly with several environmental issues like forest conservation, wild life conservation etc but no general comprehensive law existed to deal with pollution prevention, control & abatement. The Indian legal system, at present provides three major environmental laws for addressing environmental pollution. These are, (i) The Water (Prevention and Control of Pollution) Act, 1974, (ii) The Air (Prevention and Control of Pollution) Act, 1981, (iii) The Environment (Protection) Act, 1986 and Rules. The Environmental Impact Assessment Notification, 1994 and subsequent amendments in 2006, 2009, and 2011 under the EP Act are very relevant for mining activities. In this paper the provisions of the environmental laws and their applicability have been discussed.

In the Last century unprecedented deterioration in the state of Environment occurred mainly due to industrialization & Urbanization, which led to discharge of high volumes of effluents & municipal sewage to water bodies and polluted the rivers & water bodies and loss of natural habitats. Huge quantity of Industrial emissions to the atmosphere causing environmental pollution and health hazards.

Mining provides natural resources for the development of various industries and industrial processes. Mining and mineral based industries are vital to economic growth of all nations in the world. Scientific mineral survey in India was stated in 1851 by the Geological Survey of India. Geological mapping includes the search for mineral resources and the mapping has been completed for 98% of the total mappable area of the country ((SGAT, 2009). Mining industry in India has grown manifold over the years and mining operation is usually large scale and its impact persists over a long period of time..

India produces ~89 minerals including four fuel minerals, 52 non-metallic minerals, 11 metallic minerals and 22 minor minerals. India stood as world's largest producer of mica blocks & mica splitting and ranked second in the production of chromites, third in coal & lignite, and barites, fifth in iron ore, sixth in bauxite and manganese ore, eleventh in aluminum and twelfth in crude steel in the World (IBM, 1999). With its contribution of about 16 percent in exports and 20 percent in imports, the mineral sector is an important component of India's foreign trade. (Misra, 2008,).Between 1950 and 1991, mining displaced about 2.6 million people — not even 25 per cent of these displaced have been rehabilitated. About 52 per cent of these displaced were tribal. For every 1 per cent that mining contributes to India's GDP, it displaces 3-4 times more people than all the development projects put together. Forest land diversion for mining has been going up.

An estimated 1.64 lakh hectare of forest land has already been diverted for mining in the country. For instance, the forests in Burdwan have been decimated by mining. Iron ore mining in India used up 77 million ton of water in 2005-06, enough to meet the daily water needs of more than 3 million people. (Misra, 2008). Mining of major minerals generated about 1.84 billion tons of waste in 2006 and most of which has not been disposed off properly. Coal is the main mineral and every ton of coal extracted generates 3-4 tons or more of overburden.

Environmental Pollution

Mining processes such as drilling, blasting and ore extraction causes air and noise pollution. Mining causes huge land degradation and all kinds of pollution; such as air, water, soil and noise pollution. Prior to mineral extraction it involves removal of overburden (OB), its transport to the designated OB dump area and its subsequent reclamation. This activity involves air and noise pollution. Leaching of heavy metals from metal mine OB and siltation of surrounding area due to rainfall is also a serious matter of environmental concern as heavy metals can build up in the food chain.. The problem is further compounded when the ores are crushed and screened. Transportation and handling of mineral also add further to the air and noise pollution. Water pollution due to mine drainage water, (Figure 1 and 2) ore beneficiation process and tailing disposal are also essential components of mining activity, which causes significant potential water pollution. The pollution controlling mechanism is money intensive as it requires input of modern technology.

Figure 1 Mine drainage water

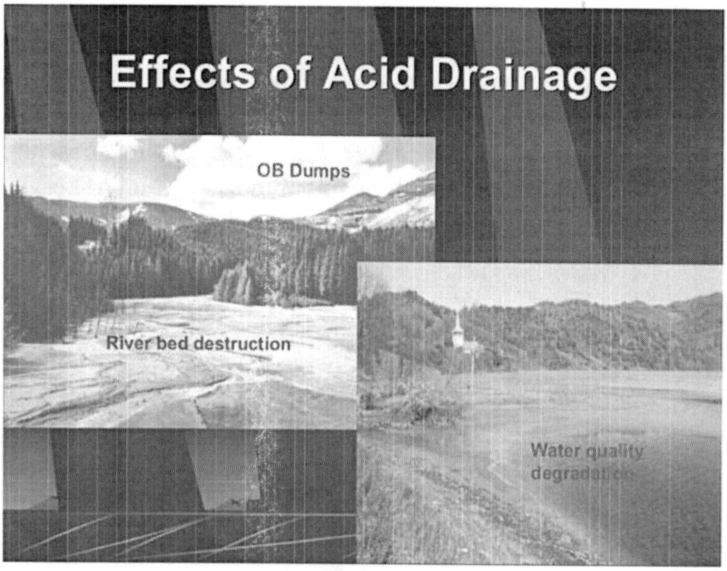

Figure 2 Effects of Acid drainage

UNO became concerned with the unprecedented environmental degradation and pollution & it was reflected at the UNO-conference on the Human Environment held in Stockholm from 5th June to 12th June, 1972, Rio conference, 1992, Review of Progress in Johannesburg,

2002.These international conferences opened a new line of thinking in different nations and every country became concerned to protect their environment, especially water quality, air quality, forests, wild life etc. India has witnessed manifold increase in Industrial sector & the increasing tendency of industrialization & urbanization is growing very fast.

In India the existing laws dealt directly or indirectly with several environmental issues like forest conservation, wild life conservation etc but no general comprehensive law existed to deal with pollution prevention, control & abatement. The Indian legal system, at present provides three major environmental laws for addressing environmental pollution. These are, (i) The Water (Prevention and Control of Pollution) Act, 1974, (ii) The Air (Prevention and Control of Pollution) Act, 1981, (iii) The Environment (Protection) Act, 1986 and Rules. The Environmental Impact Assessment Notification, 1994 and subsequent amendments in 2006 and 2009 under the EP Act are very relevant for mining activities. Besides, provisions for Public nuisance action against polluters, and the common law-right of riparian owners to unpolluted waters exist in India.

All mining operations have to comply with the environmental laws of India.

Mining activities include the following activities, which may have impact on environment. These activities are: (a) exploration, site selection, camping & construction work and consequently development of human habitation sites, township, (b) drilling, and blasting, (c) ore extraction, ore crushing, ore beneficiation, (d) transportation, (e) disposal of overburden and waste rock, (f) reclamation of voids, overburden dumps, etc, (g) tailings containment, treatment and disposal, (h) infrastructure development and access and energy source etc, and (i) mine closure.

The potential adverse impacts on environment of each of these activities comprise habitat loss and destruction, air quality, water quality, biodiversity loss, ecology and landscape deterioration, human health, resettlement and rehabilitation, social and cultural conditions etc.

The Provisions of Environmental Laws and their Applicability to Mining

Provision: The Water Act of 1974 represents one of India's first attempts to deal comprehensively with an environmental issue. The

Act was amended in 1988 to conform closely to the provisions of the Environment (Protection) Act, 1986. The Water Act is comprehensive in its coverage. It applies to streams, inland waters, subterranean waters and sea or tidal waters. Standards for the discharge of the effluent or the quality of receiving waters are not specified in the Act itself. Instead, the Act enables the CPCB & the State Boards to prescribe these standards. The Water Act regulates water pollution through a system of 'command and control. The Water Act provides for a permit system or consent procedure for operation of industries to prevent and control pollution. A person must obtain consent from the State Board before taking steps to establish any industry, operation or process, any treatment or disposal system or any extension or addition to such a system. The fact remains that 597 mines of different types (coal, Iron Ore, Bauxite, Chromites Ore, Manganese Ore, Quartz, Lime Stone, Dolomite, Sand Stone, Gem Stones, and many others) and categories (starting from 5 ha lease area to many hundreds of acres of lease area) exist in Odisha whereas only about 215 mines have registered under State Pollution Control Board under consent administration of Water Act and Air Act and from rest of the mines, some are not operating and some are operating violating legal provisions.

SPCB, Odisha does not have any mechanism to know the number of mines operating in the state .Unless the Department of Steel and Mines makes it mandatory for mines to produce clearance from SPCB, Odisha prior to allowing them to open the mines, it will not be possible for SPCB, Odisha in the present circumstances to know the total number of mines operating in the state. Besides the State Pollution Board, Odisha does not adequately monitor the mining environment with regard to water quality, air quality, hazardous waste and noise quality due to lack of man power, lack of adequate number of regional centres and adequate lab facility.

Provisions of Water Act.

The following sections of the Act are very important as they deal with different aspects of consent administration. As per the Act the following definitions are important:

Occupier- In relation to any factory or premises, means the person who has control over the affairs of the factory or the premises and includes, in relation any substance, the person in possession of the substance.

'Outlet'- includes any conduit pipe or channel, open or closed, carrying sewage or trade effluent or any other holding arrangement which causes, or is likely to cause pollution.

'Pollution'- means such contamination of water or such alteration of the (i) physical, chemical or biological properties of water or such discharge of any sewage or trade effluent or any other liquid, gaseous or solid substance into water (whether directly or indirectly) as may, or is likely to create a nuisance or render such water harmful or injurious to public health or safety or to domestic, commercial, industrial, agricultural or other legitimate use, or to the life and health of animals or plants or aquatic organisms

Sewage effluent'-means effluent from any sewerage system or sewage disposal works and includes sullage from open drains.

'Sewer'-means any conduit pipe or channel, open or closed, carrying sewage or trade effluent and 'Stream'-includes (i) River; (ii) water course (whether flowing or for the time being dry) ; (iii) Inland water (whether natural or artificial) ; (IV) subterranean waters;

(v) Sea or tidal waters to such extent or as the case may be, to such point as the state government, by notification in the Gazette, specify in this behalf;

'Trade Effluent'-includes any liquid, gaseous or solid substance, discharged from any premises used for carrying on any (industry, operation or process or treatment and disposal system), other than domestic sewage.

Some of the important provisions under the Water Act-1974 are described below:

Section – 24 : Prohibition on use of streams or wells for disposal of polluting matter

Section – 25: Application for consent, Administration of Consent by SPCB. Restrictions on new outlets and new discharges- Consent Administration by the Board

Section – 26 : Provision regarding existing discharge of sewage or trade effluent

Section -27 Refusal or withdrawal of Consent by the state board.

Section – 28: Review by an appellate authority

Section – 31 : Furnishing information

Section – 32 : Emergency measures in case of pollution of stream or well (a) removing that matter from the steam etc (NALCO-Ash pond breach case), (b) Remedying or mitigating any pollution, (c) Issuing orders immediately restraining or prohibiting the person from discharging etc

Section – 33 : Injunction from court: power of Board to make application to courts for restraining and controlling pollution of water in streams or wells.

Section – 33A : Power of the Board to give Directions- the closure, prohibition, or regulation of any industry, operation or process or (b) the stoppage or regulation of supply of electricity, water or any other service.

Sec – 41 to 50 :Penal Provisions

Section-41. Failure to comply (under section 20, 32, 33 or 33A) - shall on conviction, be punishable with imprisonment up to 3 months or with fine up to ten thousand rupees or both and in case failure continues, with an additional fine up to 5 thousand rupees for every day till the failure continues (under section 20).

Under section- 32, 33 or 33A, on conviction, be punishable not less than 1 ½ year & up to six years and with fine and in case the failure continues, with an additional fine up to five thousand rupees per day till the matter is complied. If compliance is not done for one year the prison term may extend not less than 2 years and up to 7 years.

Section-42 Penalty for certain acts.

Section -43. Penalty for contravention of provisions of section 24- Punshiable with imprisonment for a term which shall not be less than one year and six months but which may extend to six years and with fine.

Section-44. Penalty for contravention of section 25 or 26- Punishable as per the provisions in section 43 described above.

Section 45. Enhanced penalty after previous conviction- If any person who has been convicted to any offence under section 24, or 25 or 26 is again found guilty of an offence under the same provisions, he shall, on the second or every subsequent conviction be punishable with imprisonment for a term not less than two years and may be extended to seven years and with fine.

Section 49. Cognizance of offences

No court inferior to that of a Metropolitan Magistrate of a Judicial Magistrate of the First Class shall try any offence punishable under this Act. (Notwithstanding anything contained in section 29 of the code of Criminal Procedure, 1973 (2 of 1974), it shall be lawful for the Judicial Magistrate/Metropolitan Magistrate (I class) to award punishment for a term exceeding 2 years and a fine exceeding two thousand rupees).

Section 50.-Members, Officers and servants of Board to be public servants.

Sections-51-64.Miscellaneous

Section-58- Bar of jurisdiction- No civil court shall have jurisdiction to entertain any suit or proceeding in respect of any matter which an appellate authority is empowered and no injunction shall be granted by any court or other authority

Legal Implications of Non-compliance

Section 24 (i) Prohibits discharge or flow of any polluting water into any stream or well or sewer or on land and (ii) also not to impede flow of water of any stream which is likely to lead to aggravate pollution due to other causes or of its consequences.

- Exemption: If such act is done with the consent for constructions like bridge, repairs etc, for public welfare.
- The contravention of above provision is punishable under section 43 of the Act with imprisonment for a term not less than one & half year & may extend up to six years and with fine

Section 25: Restricts establishment of any industry, operation or process or any treatment and disposal system or extension or addition thereto which may discharge sewage or trade effluent into stream, well, sewer or on land without prior consent of the State Board.

Section 26: This section makes it obligatory for every existing industry to apply for and obtain consent under subsection 2 of section 25 in respect of every outlet for discharge of any sewage or trade effluent. Violation of any of the above provision i.e. section 25 & 26 is punishable under section 44 of the Act i.e. imprisonment for a term not less than two years & may extend up to six years and with fine.

Section 27: The Board has right to refuge the consent or withdraw the consent already granted under section 25 or 26 of the Act. It can also be for noncompliance of the conditions imposed under the consent.

Section 47: Offences by Companies:- Every person, who at the time of the offence committed was in charge and was responsible to the company, for conduct of business of the company, as well as the company shall be deemed to be guilty of the offence and shall be liable to be proceeded against provided that nothing will render such person liable to punishment if he proves that the offence was committed without his knowledge or that he exercised all due diligence to prevent the commission of such offence. If it is proved that the offence has been committed with the consent or connivance of or is attributable to any neglect on the part of the Director, Manager, Secretary or other officer of the company, shall be deemed to be guilty and liable to be prosecuted and punished

Section 33A: Authorizes Board to issue directions in writing to any person, office or authority who shall be bound to comply with such directions:

Power to issue directions includes to direct – (a) the closure, prohibition or regulation of any industry, operation or process; or (b) the stoppage or regulation of supply of electricity, water or any other serviced

Section 45A provides that whoever contravenes any provision of the Act or fails to comply with any direction given there under for which no specific penalty has been provided shall be punishable with imprisonment of three months and also a fine up to Rs.10, 000/- or both and if contravention or failure continues, a further fine of Rs.5000/ - for each day of such continuance

Section 42 prescribes penalty for any act contrary to the provisions of the Act or intervenes with the property of the state Board or meter or gauze or other measuring and monitoring devices will be punishable with the imprisonment of three months or with fine up to Rs.10, 000/ - or both.

The Air Act 1981

The Air Act was enacted in 1981 under Article 253 of the Constitution of India, which gives power to the Central Government to make laws to implement decisions taken at International Conferences.

The Act represents the implementation of the decisions made at the United Nations Conference on the Human Environment held at Stockholm in 1972.

The Air Act provides power to State boards for the protection of environment

With increasing industrialization and tendency of cluster /aggregate in few areas, the problem of air pollution has been felt. The problem is acute in densely populated areas.

The presence of various pollutants discharged through industrial emissions and from certain human activities connected with traffic, domestic fuel use, heating, incineration, refuse etc has detrimental effect on health of people and on biodiversity.

Air does not have boundaries; pollutants released at one place are carried to distant places and create health problems. Some definition and provisions are described below.

Definitions

'Air pollutant' means any solid, liquid or gaseous substance (including noise) present in the atmosphere in such concentration as hay be or tend to be injurious to human beings or other living creatures or plants or property or environment.

Some of the provisions of the Act are:

- Section- 18 : Power to give directions
- Section – 19 : Declaration of restricted areas Section- 20 : Powers to give instructions for ensuring standards for emission from automobiles
- Section – 21 : Restrictions on use of certain industrial plants

Section - 22 : Industries not to allow emission of air pollutants beyond the standard lay down by the state board.

- Section - 22A : Power of board to move the court for restricting from causing air pollution by industry / persons
- Section – 24 : Power of Entry by the officials of the Board
- Section – 25 : Power to obtain information
- Section- 26 : Power to take samples of Air or Emissions & procedures
- Section-31 : Appeals-Appellate Authority
- Section- 31A: Power to give direction for closure, prohibition or regulation of any industry

- Sec – 37 to 46: Penal Provisions
- Sec- 47-54 : Miscellaneous

Legal Implications of Noncompliance:

Section 21: Provides that without previous consent of the state Board no person shall establish or operate any industrial plant in an air pollution control area.

Note: In Odisha the entire state has been declared as air pollution control area.

Section 22: No person operating any industrial plant in any air pollution control area shall discharge or cause or permit to be discharged emission of any air pollutant in excess of the standards laid down by the state Board

Section 31A: Authorizes to the Board to issue directions in writing to any person/officer or authority that shall be bound to comply with such directions. The Board to issue directions (a) the closure, prohibition or regulation of any industry, operation or process; or (b) the stoppage or regulation of supply of electricity, water or any other service.

Failure to comply with the provisions has been made punishable under section 37 of the Act

Section 37. provides that person failing to comply with the provisions of section 21 or 22 or 31, shall, in respect of each such failure, be punishable with the imprisonment which shall not be less than for a term of one year and six months but may extend to six years and fine and in case the failure continues an additional fine of Rs.5000/ - for every day of such continuance after conviction of first such failure. It is further provided that if contravention continues beyond a period of one year after the date of conviction, offender shall be punishable for an imprisonment of term not less than two years and be extended to seven years with fine

Section 38 & 39: Provides for any other violation of the provisions of Act or on failure to intimate the occurrence of the emission of air pollutants into the atmosphere in excess of the standards or apprehension of such occurrence or giving any false statement in any material particular for obtaining consent u/s 21 or giving any other false information to Board has been made punishable with the imprisonment for a term which may extend to three months or fine up to Rs.10, 000/- 0r both.

Section 40: contains the provision to deal with offences of companies which are similar to section 47 of the water pollution Act

The Environment Protection Act-1986.

In the wake of the Bhopal Gas Tragedy in 1984, the Government of India enacted the Environment (Protection) Act of 1986.The Act is an umbrella legislation designed to provide a framework for Central Government coordination of the various Central and State authorities established under the Water (Prevention and Control of Pollution) Act, 1974 and Air (Prevention and Control of Pollution) Act, 1981. The EPACT envisages filling up gaps in the legislative protection to the environment. The main thrust of the Act is to make top management responsible for compliance with provisions of Environmental laws: Liability & prosecution; Statutory provision to empower executive to issue notification, directions, standards and regulations for compliance to meet the required contingencies; Private citizens empowered to file cases against violation of environmental laws against industries after notice to authority; Punishments and penalties made more stringent; Powers conferred on enforcement agencies to effect closure of polluting industries; By 1992 amendment- industries have to furnish annual environment statement to regulatory bodies and observe self discipline. Some of the provisions under the Act are:

Section – 2: Deals with the definition of different terms:

Environment- Includes water, air and land and the interrelationship which exists between water, air, land and human beings and other living creatures, plants, micro-organisms and property. [Section 2 (b)]

Environmental Pollutants:

Any solid, liquid, or gaseous substance present in such concentration as may be or may tend to be injurious to the environment. [Section2 (c)]

Hazardous substances

Hazardous substance - Any substance or preparation which by reason of it's chemical or Physico-chemical properties or handling is liable to cause harm to human beings, other living creatures, plants, micro-organisms, property or the environment. [Section2 (e)]

Occupier

Occupier- In relation to any factory or premises, an occupier means a person who has control over the affairs of the factory or the premises and includes in relation to any substance, the person in possession of the substances.

Section 5: ***Power to issue directions*** *(similar to section 33A of Water Act-1974, section-31A of Air Act-1981).* Under Section 5, the Central Government is empowered to issue directions in writing to any person, officer, or any authority and such person, officer or authority shall be bound to comply with such directions. The power to give directions under this section includes the power to direct (a) the closure, prohibition or regulation of any industry, operation or process; or (b) stoppage or regulation of the supply of electricity or water or any other service. The power is delegated to State Government/ SPCB.

This is an important section as it empowers government to give directions for preservation of ecological order and social harmony and pollution free environment and bring awareness on environmental conservation through media and press etc. (Reference may be made to (i) Dahanu Taluka Environment Protection Group vs. Bombay Suburban Electricity Supply Company Ltd (1991) -2.S.C.C.539.

The Supreme Court of India gave direction to make environment a compulsory subject in schools and colleges in a graded system (M C Mehta vs. Union of India A.I.R. 1992 S.C.382)

Section 6: *Rules to regulate environmental pollution.*

Ministry of Environment and Forests (MoEF) is responsible for making rules to implement the Environment Protection Act (EPA) and has adopted industry specific standards for effluent discharge and emissions for 24 designated industries.

Section 7: This section expressly states that those persons who are carrying on industry, operations etc., are not to allow emission or discharge of environmental pollutants in excess of the standards

Section-8 : Persons handling hazardous substances to comply procedure

Section 9 : *Furnishing of information in certain cases*

In cases where discharge of an environmental pollutant occurs in excess of the prescribed standards, or there is an apprehension of this occurring due to an accident or some other unforeseen event, then the person who is in charge at that particular time is responsible for preventing environmental pollution due to the said discharge.

It is the responsibility of the person to intimate the occurrence of the discharge to the CPCB/SPCB, and provide all possible assistance to the regulatory agencies to counter the ill effects of the discharge. All expenses incurred by the regulatory agency in undertaking remedial measures may be recovered from the polluter. In *Indian Council for Environ-legal Action vs. Union of India* (Supreme Court, 1996), the court held that the Central Government has the power under Sections 3 and 5 of the EPA to levy and recover the cost of the remedial measures. The polluter pays principle was also held to be applicable and Sections 3 and 5 of the EPA were held to empower the Central Government to give directions and take measures for giving effect to this principle.

Section 10: *Power of search and entry*

The E.P. Act empowers authorized officers to enter, at all reasonable times, any place, or premises for determining the infringement of standards and procedures set under the Act, and for testing any equipment, industrial plant, record, register or document or any other material object for determining offences under the EPA.

A willful delay or an obstruction of such an inspection is an offence and the person in-charge is liable for prosecution

Section 15: *Penal provision*

EPA provides for punishment with imprisonment for a term which may extent to five years or with a fine which may extent to one lakh rupees, or with both.

In case the contravention continues, there may be an additional fine of Rs 5000 for each day of the contravention.

If the contravention continues beyond a period of one year after the date of conviction, the offender is punishable with imprisonment for a term which may extent to seven years

Section 16: *Offence by a company*

When an offence under this Act has been committed by a company, then every person directly in charge of and responsible to the company at the time of the offence, shall be deemed to be guilty of the offence, unless he or she proves that the offence was committed without their knowledge **or** that they exercised due diligence to prevent the offence.

Section 17: *Offence by Government department*

In case a Government department has committed an offence under this Act, then the head of the department is deemed guilty unless he or she proves that offence was committed without their knowledge or that they exercised due diligence to prevent the offence. Any other person other than the head of the department, who has consented to the offence or was in connivance, or neglected their duties in preventing the offence, shall also be guilty under the Act.

Section 18: *Protection of action done in good faith*. No suit, prosecution, or other legal proceedings lie against the Government or any officer or employee of the Government or any authority constituted under this Act, or any member, officer or employee of such an authority, in respect of anything which is done in good faith in pursuance of this Act, Rules made under the Act and orders and directions issued under this Act.

Section 19: *Cognizance of offences (citizen's suit)*

A court can take cognizance of an offence only on a complaint made by the Central Government, or any authority or officer authorized or by any person who has given notice of not less than 60 days to the Central Government, or any authority or officer authorized.

This citizen's suit section appears to give significant powers to the citizens to enforce the EPA. However, under the EPA, only Government officers have the power of sample collection, which may be adduced as evidence during proceedings. This particular aspect seems to weaken the Section

Section 22: ***Bar of jurisdiction***

No civil court has the jurisdiction to entertain any suit, or proceedings in respect of anything done, action taken, or order or direction issued by the Central Government or any other Authority or officer in pursuance of any power conferred by or in relation to its or his /her function under the Act

Section 23: *Powers to delegate*

The Central Government can delegate its functions (as specified in Section 3 of the Act).

Section 24: Effect of other laws- (i) Subject to the provisions of sub-section (2), the provisions of this Act and rules or orders made therein shall have effect not withstanding anything inconsistent therewith contained in any enactment other than this act, (2) Where any act or omission constitutes an offence punishable under this Act and also under any other Act then the offender found guilty of such offence shall be liable to be punshied under the other Act and not under this Act.

Section-25. Power to make rules-The Central Government may, by notification in the Official Gazette make rules for carrying out the purpose of this Act. (2) In particular, and without prejudice to the generality of the foregoing power, such rules may provide for all or any of the ten following matters (a) the standards for discharge and emissions of environmental pollutants under section 7; (b) procedure for handling of hazardous substances substances under section 8; (c) matters of sub-section (1) of section 9; (d) manner of taking samples of air, water, soil or other substance for analysis as dealt with under sub-section (1) of section 11; (e) manner of serving the analysis report under clause (a) of sub-section (3) of section 11; (f) the functions of environmental laboratories, procedure of submission of samples to those laboratories, form of laboratory report, fees payable for such report and other matters for good functioning of laboratories under sub-section (2) of section 12; (g) the qualification of government Analyst as under section13; (h) the manner of serving the notice of the offence and intention to make a complaint to the Central Government as under clause (b) of section19; (i) the authority or officer to whom any report, statistics, returns, accounts and other information shall be furnished as under section 20; (j) any other matter which is required or may be prescribed by the Central Government.

26. Rules made under this Act to be laid before Parliament. The procedure and time period have been prescribed.

Additional Provisions Applicable to Mining: Environment (Protection) Act, 1986 and Rules 1986

(Notifications issued under the Act).Two important notifications applicable to mining industries are listed below:

- Coastal Regulation Zone Notification (CRZ) (1991)
- The Environment Impact Assessment (EIA) Notification, (1994) and as amended in (1997), 2006, 2009 and 2011.

- Note: Under EIA Notification, 1994, section 2 stipulates requirement of prior Environmental Clearance (EC) for all new projects or activities listed in the Schedule to this notification and for Expansion & modernization of existing projects listed in the Schedule to this notification. This is again reiterated in section 7of EIA Notification.

Environmental Impact Assessment, (EIA) 1994

Until 27, January 1994, obtaining environmental clearance from MOEF was only an administrative requirement intended for mega projects undertaken by the Government or public sector undertakings. The EIA notification of 14, Sept.2006, makes EIA statutory for 40 items (A & B category) under 8 projects/Activities (Schedule of the notification).There will be a Central Expert appraisal Committee for all projects included under MOEF (A- category) & a state level EIA Authority for B category projects .

Further Amendments in 1997, 2006, 2009and 2011 (Coastal Regulatory Zone (CRZ) matters) : The salient provisions of the notification are as follows: All projects listed under Schedule are required to obtain environmental clearance (EC) from the MoEF/State Government. Projects that fall under 'A' category need clearance from the MoEF.

All development projects whether under Schedule or not, if they are to be located in certain notified ecologically sensitive or fragile areas will have to obtain clearance from MoEF. Industrial projects are further required to obtain a LOI (letter of intent) from the Ministry of Industry, and a NOC (no objection certificate) from the SPCB and the State Forest Department if the location involves forestland. Once the NOC is obtained, the LOI is converted to an industrial license by the State authorities. However, if the project falls under Schedule of the EIA notification, (A-category) it must obtain an environmental clearance from the MOEF.

In all cases an application seeking prior EC shall be made in the prescribed Form-1 and with a pre-feasibility project report except that, in case of construction projects or activities in addition to Form-1 and the supplementary Form-1A, a copy of the conceptual plan shall be provided instead of the pre-feasibility report. Form-1A includes the checklist of environmental impacts such as (i) Land Environment; (2)

Water Environment; (3) Vegetation) ; (4) Fauna; (5) Air Environment; (6) Aesthetics; (7) Socio-Economic Aspects; (8) Building Materials; (9) Energy Conservation; (10) Environment Management Plan.

Environmental Clearance (EC) will have four stages (i) Screening, (for A, B1& B2 categories), (ii) Scoping for A and B1 category of projects, (iii) Public Consultation (for A and B1 categories- except building, township etc, Defense, Expansion of Roads & Highways, modernization of irrigation projects & B2 category projects), (iv) Appraisal (detailed scrutiny by EAC or State level EAC).

The EIA report that is submitted after the site clearance contains the following : (i) Description of site, site map, (ii) Land requirements, No objection certificate from State Forest Department if forest land is involved, (iii) Rehabilitation plan for displaced people, (iv) Consent to establish from SPCB (regarding effluents and emissions), (v) Report on impacts on the environment during construction, (vi) Collection of ambient air quality and meteorological data, (vii) Report on hydrology and water quality, (viii) Report on occupational safety and health, (ix) Details of transport and handling of raw material, (x) Report of impact on sensitive terrestrial ecosystems.

EIA generic structure includes the following:

1. Introduction-purpose of the report, identification of project and project proponent, brief description, scope of study as per term of reference (TOR)) 2. Project description, 3. Description of the Environment-study area, components & methodology, baseline for valued environmental components as identified in the scope, Base maps of all environmental components, 4. Anticipated Environmental Impacts and Mitigation Measures, 5.Analysis of Alternatives (Technology & Site), 6. Environmental Monitoring Programme, 7. Additional studies (Public Consultation, Risk assessment, Social Impact Assessment, R& R Action Plans, 8.Project Benefits, 9.Environmental Cost-Benefit Analysis, 10. EMP, 11.Summary & Conclusion (This will constitute the summary of the EIA report), 12. Disclosure of Consultants engaged.

The EIA document is designed (i) to impose certain restrictions and prohibitions on some projects or activities and expansion and modernization of existing capacity of projects/ activities, (ii) to seek public consultation from stakeholders, (iii) to incorporate necessary environmental safeguards through Environ Management Programme by assessing the impacts at planning stage itself through transparent process involving consultation of public and relevant experts.

The general conditions of EIA notification stipulates that (i) any project or activity specified in category B will be treated as category-A, if located or in part within 10km from the boundary of (a) protected areas notified under the Wild Life (Protection) Act-1972; (b) critically polluted areas as notified by CPCB from time to time; (c) notified ecosensitive areas; (d) interstate boundary and international boundaries

EP ACT: A Critique

(1) Section 19 (Citizen's suit provision) of the Act seemingly gives the public significant powers to enforce the Act. However, only Government officials are given the power under the Act to collect samples needed as evidence. In addition, during the sixty days notice period required for the Government to decide whether to proceed against the alleged violation, the offending industry has time to clean up and prepare itself for collection of samples. Sixty days is very long period.

(2) Under the Section 24 of the Environment (Protection) Act 1986, in case of a violation both under the EPA and Water Act or Air Act, the penal provisions of the Water Act or the Air Act would usually apply. The high penalties thus provided in the E.P.Act-1986 are therefore illusionary for those violations where Water Act and Air Act also apply.

(3) Research Foundation for Science Technology and National Resource Policy Versus

Union of India

Selected Local residents should be appointed as wardens for environmental surveillance, particularly to take note of illegal dumping of hazardous wastes. Access to public records with the environment protection authorities should be freely allowed to the public, as the right to a healthy environment has been defined as part of the Right to Life under Article 21 of the Constitution. Relevant important information should be displayed on notice boards and newspapers and communicated through radio, television and the Internet. The Court decisions stipulate that all industries, involved in hazardous chemicals and generating hazardous wastes display on-Line date outside the factory gate, on quantity and nature of hazardous chemicals being used in the plant, as well as water and air emissions and solid wastes generated within the factory premises. If such data is not made available, the unit should be asked to show cause or even be asked to close down.

Relevant information should be displayed on notice boards and newspapers and communicated through radio, television and the Internet .Besides the informers and "whistle-blowers" within industry, who provide information, should be protected and strict confidentiality about them maintained.

Third-party audit of hazardous wastes, where the audit team includes members of the community, should be made a routine practice." This is very welcome directions of the court.

(4) Under EIA and EMP the flora (vegetation) and fauna enumeration and quantification have not been adequately defined. The forest departments may not have the adequate expertise to enumerate and quantify the flora (tree, shrub, herb etc). The wildlife departments of the government usually study the large-sized fauna such as mammals, birds, and reptiles. They do not enumerate invertebrates and soil fauna, which are ecologically very important. University-biology faculty and individual biodiversity experts should be involved. The enumeration profile and level of quantification should be specified. This requires detail study in immediate post-rains and in summer, by experts. Since the rural livelihood is biodiversity based and biodiversity conservation is a priority area as per United Nations mandate, this aspect should get more scientific attention.

Regulatory Provisions of Water and Air Acts:

From environmental point of view, the mines are regulated by SPCB using consent to operate under both the acts as an instrument. Consent to operate is granted under section 25/26 of the Water Act-1974 and under section 21 of the Air Act-1981 on pollution control matter, which primarily specify the quality and quantity of wastes/ pollutants generated in mining operations, method of management of effluent, emissions, permissible limit of effluent and emission discharges, and control methods, solid waste generation and management as per law and eco-friendly manner.

During 2006-2007, SPCB, Odisha made a special drive to bring all mines into the consent administration of the Board. After obtaining the list of mines from the Directorate of Mines, Government of Odisha, the defaulting operating mines were identified and follow up action was taken by the SPCB to bring them into consent administration.

Consent to operate is generally granted after evaluating the steps taken by the concerned mine authorities and evaluating the status of implementation of various pollution control measures adopted by the existing mines.

Besides status of implementation of the following issues are also considered: (i) whether wet drilling and controlled blasting is adopted to prevent generation of fly dust and fly rocks? ; (ii) whether water sprinkling by automatic sprinklers or by manual method or by mechanized methods, is done on haulage roads to prevent generation of fugitive dust? (iii) Whether overburden is dumped at appropriate place and is stabilized by plantation?;

(iv) Whether garland drain communicating to a settling pond is provided around the OB dump for controlled discharge of surface run off? (v) whether check dams at appropriate places of the mine has been constructed and are in working condition to prevent direct disposal of surface run off ?; (vi) whether mineral processing plants with crushers and screens are operating with the mine lease area ? And if so, whether requisite air pollution control devices are in operation?; (vii) whether mine drainage/waste water and ambient air quality remain within the prescribed standards ? .The operation continuance or closure of mining activities will depend upon satisfactory compliance of these aspects .

Reclamation and rehabilitation of mine: Reclamation and rehabilitation issues are regulated under Mineral Conservation and Development Rules (MCDR), 1988 and rule was enacted in 2003. As per rule 23A of MCDR, 1988, every mine has to prepare progressive mine closure plan at the time of fresh grant or renewal of lease and one year prior to the proposed closure of mine respectively. This provision is mandatory. This plan shall be approved by the Regional Controller of Mines, IBM, or by the office authorized by the state government.

Besides as per Rule 23F of MCDR, financial assurance along with mine closure plan and procedure for forfeiting the surety amount in case of failure to carry out reclamation and rehabilitation work exist.

Regulatory Provisions of Environment (Protection) Act, 1986 for the opening of New Mines:

The environmental aspects of mining are regulated by the MoEF, Government of India using Environmental Clearance (EC) as instrument under the Environment (Protection) Act-1986. The liability of obtaining EC has been stipulated by EIA notification-1994 under the following circumstances, (i) New Mining projects (major minerals) with mining lease area more than 5 hectare (ha) ; (ii) Expansion of production.

MoEF in their 25 April 2005 notification clarified the liability to obtain EC under the following circumstances, (i) Mining projects, which continue to operate at pre-1994 level or less or did not require renewal so far after the notification (s) coming into force, shall obtain EC at the time of renewal as and when due; (ii) Those mining projects, which have expanded either in terms of lease area or production after the notification (s) coming into force even it is as per approved mining plan or have got their lease renewed after 1994 but have not obtained EC so far shall obtain EC; (iii) Those mining projects, which have obtained EC under the provisions of EIA notification, 1994 will not require EC at the time of renewal if no expansion, either in the form of production or lease area has not taken place.

In EIA notification of September, 2006, both major and minor mineral were included under the purview of obtaining EC and the liability of obtaining EC was stipulated as follows: (i) New mining project (major or minor minerals) with lease area of 5 ha or more; (ii) Expansion of existing project; (iii) At the time of renewal of lease. Mining leases of 5 ha to less than 50 ha came under the State Environment Impact Assessment Authority and mining leases of 50 ha and more than 50 ha came under the control of MoEF, New Delhi. The SPCBs are pursuing with all mining units to take action for compliance of the provisions of the statute.

Violation of the provisions of EIA notification entails action under EP Act-1986. It has been notified that MoEF, Government of India and Environment Departments of State governments are the appropriate authorities for taking action under section-19 of EP Act-1986.Besides; the Hazardous Waste Management Rules-1989 also applies to mines generating hazardous substances. The hazardous waste includes only the quantity of oil and its storage in mining lease area and the used oil etc by mines .These mines are to take Hazardous Waste Authorization from SPCB to operate. At present 77 mines in Odisha have taken authorization for hazardous waste management from SPCB, Odisha (personal information from SPCB). These mine types are: 20 coal, 1 lime stone, 4 Dolomite and lime stone, 20 Iron ore, 12 Iron and manganese ore, 8 Manganese, 9 chromites, 1 Iron and dolomite, 1 Bauxite and 1 Iron and Bauxite .

Acknowledgement

I wish to thank Dr.S.K.Sarangi, SGAT, for motivating me to write this paper. I wish to thank Mr.Chitaranjan Nayak, senior Environmental Scientist, Er.Simachal Dash, Environmental Engineer of SPCB, and Odisha for providing data on the number of mines operating in the state and for discussion and Er.S.K.Guru of AdityaBirla Aluminum Project, Rengali for providing some photographs.

References

Dash, M.C. (Ed)., 2006. The State of Environment-Orissa, State Pollution Control Board, Orissa, Bhubaneswar.

IBM, 1999. Annual Report.

Misra, S.2008. Sustainable Mining, In: Dash, M.C. and Acharya, K.C. (Ed), Environmental Management for Sustainable Development in Industries, 62-66.Interscience Institute of Management & Technology, Bhubaneswar.

Pollutions Control Acts, Rules, and Notifications Issued there under, 2006. Central Pollution Control Board, 2006.

Vision Document on Mineral Development-2020 for Orissa. 2009. Society of Geoscientists and Allied Technologists (SGAT), Bhubaneswar.

The Environment (Protection) Act, 1986, Professional's., 2006.Professional Book Publishers, New Delhi.

The Water (Prevention and Control of Pollution) Act, 1974. Universal's Bare Act with short notes, Universal Law Publishing Co.Pvt.Ltd, Delhi

SPCB Officials (Er.N.Sahoo, Mr. C.R.Nayak, Er.Simachal Dash) -Personal Communication

20

Industrial Impact on Skin and Respiratory System: A Case Study of Health Hazards In Balasore, Orissa*

Introduction

The post independence period has witnessed considerable growth of industries in the country. There is shifting of the population from rural to urban areas. People are forced to get out of the safe hands of nature (Chakrabarty, 1967). The resultant industrial activity is associated with problems of the environmental pollution. The industrial wastes and effluents, emissions, wide spread use of insecticides, synthetic detergents, inorganic nitrogen fertilizers, etc. have recoiled on human beings and contaminated the supply of essentials like water and air. So the environmental pollution arising out of industrialization coupled with human factors adds associated health hazards.

The arena of study comprises industrial estates with metallurgy, rubber, plastic and paper industry. The various hazardous waste

---

* Choudhury Satyabrata. Nanda and Madhab Chandra Dash

generated by these industries produces cumulative toxic effects over time with impact on the health of the population of the villages around industries (Fig. 1).

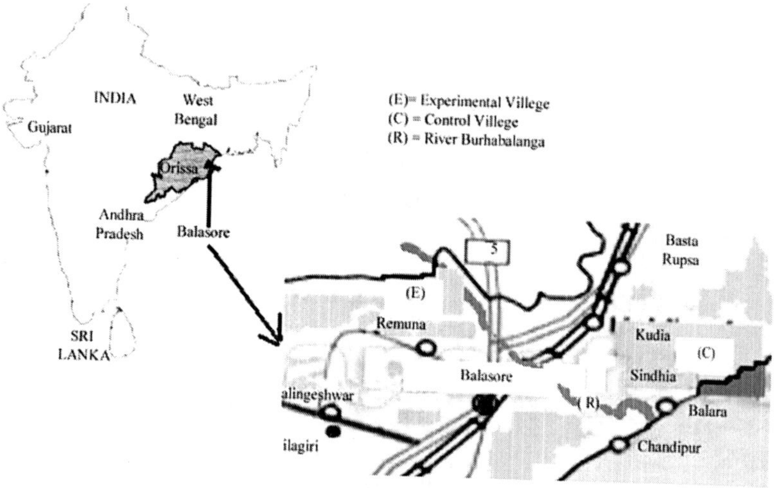

Figure 1: Location of study area

This prompted to measure the health status of the population around industrial estate, workers of the industry and to quantify their health problems and to test the hypothesis that the skin and respiratory diseases are more prevalent in the population. The work forms a part of D. Sc. thesis submitted in F.M. University, Balasore by the first author (C. S. N.) under supervision of the second author (M. C. Dash). The present communication deals with a case study of health hazards particularly the effect on skin and respiratory system due to industrial waste in Balasure of Orissa, India.

Material and Methods

A retrospective cross-sectional case-control study technique was adopted to test the causal hypothesis basing on the distinct features like: (a) Both exposure and outcome (disease) have occurred before the onset of the study (b) The study proceeds backwards from effect to cause and (c) It uses a control or comparison group to support or refute an inference.

Twelve villages with a total population of 14, 346 were selected for case studies (experimental group), located surrounding the industries i.e. Birla Tyres, Balasore Alloys, Emami Paper and Ori-Plast.

Table 1: Percentage of diseases among population under study

Disease	Percentage of people suffering in experimental group (villages around industries) (n=6298)				Percentage of people suffering in control group (villages away from industries) (n=399)				Percentage of Industrial workers suffering from diseases (n=458)		
	M	**F**	**C**	**Total**	**M**	**F**	**C**	**Total**	**M**	**F**	**Total**
Skin	3.51	3.12	0.59	7.22	1.25	1.65	0.60	3.50	6.13	1.12	7.25
Respiratory	2.10	2.80	1.46	6.36	1.66	1.13	0.86	3.85	15.73	1.39	17.12

M- Male, F- Female, C- Children

Two villages with a total population of 2044 were for study of comparison (control group) which was located twenty five Kms away from the industry separated from experimental group by the river Burhabalanga (Map 1). The questionnaire survey of both groups was conducted by four skilled workers trained by the researcher with periodic supervision of the field work. Clinical study was carried out at selected nodal points by a medical expert team with a respondent of 6298 in experimental and 399 in control group. The relevant data from the operational health care units of the area and dispensaries of the industry were obtained.

Results

Clinical survey revealed Skin and Respiratory problems among the population of villages around the industries and workers of different industries under study. Possible determinants for the health hazard were identified. The prevalence of skin diseases (7.22%), and respiratory problem (6.36%) were in experimental group (Industrial villages) and skin disease (3.50%), respiratory (3.85%) were in control group (Non Industrial villages). Occurrence of disease in male, female and children distributions were studied (Table 1).

The prevalence of different diseases among workers of the industries under survey was skin disease (7.25%) and respiratory disease (17.12%). Data from socio-economic profile revealed 10.66% of population of the industrial villages work in the industries under survey.

Data collected from questionnaire survey, clinical study and operational health care systems were transformed into information by the application of statistical analysis technique. The different statistical analysis revealed the prevalence of diseases were more in experimental group (Industrial villages) than the control group (Non Industrial villages) as evident from normal Proportion test ('z' test), Chi-square test and student 't' test (Table 2). The association between occurrence of diseases and distance of village from Industry were tested by correlation coefficient analysis. It was evident from the analysis that diseases were negatively correlated establishing a strong association between the occurrence of disease with the distance of villages from industries (Fig. 1).

Table 2: The results of different statistical analysis

Different Test	Results
'z' Test	Small proportion of the people in control and high proportion of people in experimental group are affected by the disease ($p<0.05$).
'χ^2'	Experimental group is affected by most of the diseases ($p<0.05$).
't' Test	There is difference between the average numbers of people affected by the distance in experimental and control group. But the difference is not statistically significant at 0.05 level.
Correlation (r)	Diseases are negatively correlated indicating strong association between the occurrence of diseases and distance of villages from industries i.e. more the distance from industrial estate less will be the occurrence of diseases.

Discussion

Skin diseases

In experimental group, total numbers of 384 (7.22%) persons were suffering from skin diseases out of which 3.51% males, 3.12% females and 0.59% was children. In the control group, total numbers of 46 (3.50) persons were suffering from skin diseases out of which 1.25% males, 1.65% females and 0.60% were children (Table 1). In experimental group 6.12% were suffering from eczematous dermatitis and the rest included bacterial and fungal diseases. In control group 3% were suffering from parasitic skin disease like scabies, fungal and rest suffering from allergic dermatitis. In the present study, percentages of Industrial workers suffering from skin disease were 7.25% out of which 6.13% were males and 1.12% females.

The skin problems were detected to be allergic in nature. The higher proportion of skin diseases among population towards industrial estate and gradually declining while away from industry raises the suspicion of higher concentration of allergen around the industrial areas which was responsible in the causation of disease. Again a higher proportion of skin disease among industrial workers suggests a consistency in the causal hypothesis and a strong association of the industrial impact.

In the present study, the nature of wastes by different types of industries reveal used oils, spent chemicals, corrosive waste arising from strong acids and bases are generated from Paper industries. Rubber industry generate oily sludge, hazardous chemicals from discarded container, scrap compound containing aromatic, napthamic, paraffin oil, etc. Metallurgical industry generates used oil, flu gas cleaning, used batteries, suspended particles, slag etc and Plastic industry generating PVC scraps containing residues and stabilizer etc. All these hazardous wastes are discharged to outside through different channels like Sapna nallah which ultimately drain to Sono river. Some are staked inside the factory premises, effluents also mix with rain water and increases the chances of leaching which further contaminate near by water bodies. The waste water generated from leakages also contaminates soil and ground water.

Dash, (2004) discussed hazardous waste as corrosive, ignitable, reactive or toxic or capable of posing substantial threats to human health and the environment and is to be treated separately from all other waste.

A study conducted on (Brahma et al., 2003) pollution in shallow aquifers in and around industrial estates near Balasore, Orissa reveals incidences of high contents of physio-chemical parameters in ground waters. High concentrations of some heavy metals like nickel, chromium and cadmium and chemical constituents in the effluents in few industries were found exceeding their respective tolerance limit for its discharge into inland surface water.

Allergic eczema (Banerjee and Banerjee, 2009) is most often caused by nickel, rubber, metals and gloves. Important cause of allergic eczema in women were nickel, other metals, rubber and gloves and in men rubber, nickel, gloves, wood products and chromium in that order. 50% of all eczematous diseases were caused by detergents, water, metals food stuffs and rubber.

The metal dermatitis is common near industrial areas (Clark, 1992). Nickel is the most common sensitizer around industrial areas and also affecting the general population, specifically among females due to early exposure to nickel following the practice of ear piercing. Chromate is one of the highest contactant in the occupation relating setting. Due to ubiquitous nature of chromates found in cement, leather, the dermatitis runs a chronic course and also it is frequently relapsing. It is interesting to note that females shows a high incidence of hand eczema which occurs due to rubber and chromium.

Rubber products related industries have high incidence of allergic contact dermatitis to rubber products (Philipson, 1984) particularly posing the occupational hazards. Acids, bases and oils liberated from industries as waste also causes allergic dermatitis.

Waste water of industries contains heavy metals like zinc, chromium, lead etc above the permeable limits alters the nutritional value of vegetables grown, and long run consumption of such vegetables may induce health hazards in human beings which are a matter of concern (Marwari et al., 2009).

Zinc compounds are astringents, corrosive to skin, eye and mucus membrane. They cause a special type of dermatitis known as 'Zinc Pox'. Therefore it is desirable to treat such water to bring the level of heavy metals to acceptable limit before its use (Ray and Khangarot, 1989).

Respiratory diseases

Industrialization has raised a significant number of respiratory illnesses like pulmonary tuberculosis and pollution related illness like asthma and bronchitis.

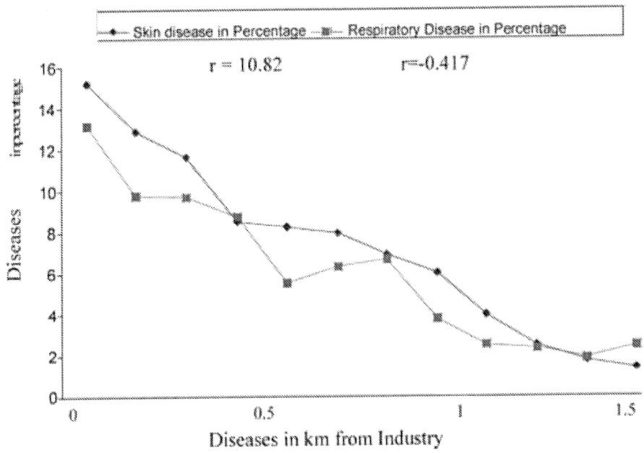

Figure 2: Relationship between diseases and distance

The occurrence of respiratory problem in experimental group, recorded to be, 6.36% out of which 2.10% males, 2.80% females and 1.46% were children. In the control group, total numbers of 3.85% persons suffering from respiratory diseases were recorded out of which 1.66% males, 1.13% females and 0.86% was children. Percentages of Industrial workers suffering from respiratory infection were to be 17.12% out of which 15.73% were males and 1.39% females. Asthma like symptoms were noted among all the workers reported to be suffering from respiratory illnesses.

Cases of non tubercular respiratory complication were bronchitis, allergic bronchitis and bronchial asthma. In experimental group, 94.17% cases and in control group, 57.14% were reported to be suffering from allergic bronchitis.

Spirometry study was conducted over the cases of non tubercular respiratory illnesses. 12.94% were found to be of restrictive variety and 86.06% were of obstructive in experimental group. In control group, all the cases were of obstructive type.

Non tubercular respiratory diseases are a group of conditions like allergic bronchitis, and allergic asthma. It also accommodates work place–related asthma, which is a growing problem accounting upto 15% of all cases of asthma (Donham, 1982). Work related asthma is classified as either occupational or work aggravated. This is a condition of variable degree of air way obstruction or bronchial hyper responsiveness caused by agents or conditions in the work place. Various allergens in the form of pollutants aggravate asthma like symptoms (Jarup, 1996).

Gases, vapours, fumes, smokes including chlorine, nitrogen dioxide, sulphur dioxide and ozone are found as pollutants in the work place and out side work place of industrial areas which precipitates asthma like symptoms (Anya and Kay, 1994).

India is a populous country in the world. Urbanization and industrialization has raised a significant number of pollution related respiratory illness like asthma and bronchitis. The main pollutants are suspended particulate matter (SPM), nitrogen dioxide, sulphur dioxide and carbon monoxide. SPM normally cause allergies, respiratory illnesses and eye irritation. The lead content in SPM from vehicle exhaust leads to cell injuries and affects the central nervous system. In a report published by Phadtare et al., (2005), SPM levels exceeds quality guidelines of World Health Organization (WHO) in more than 98% in industrial areas and 90% in metropolitan cities as discussed by a World Bank Report. Nitrogen dioxide and sulphur dioxide cause inflammation of the respiratory track, making people more vulnerable to the effects of other allergens and viral infection (Banarjee, 1985).

The higher percentage of respiratory illness among the industrial workers and population residing close to industrial estate (experimental group) may be the out come of interaction of Skin disease in Percentage Respiratory Disease in Percentage

Diseases in km from Industry the impact of major pollutant like Suspended Particulate Matter (SPM) and other pollutants coupled with other minor factors like poor living conditions, malnutrition and over crowding.

The national survey reported a higher incidence of tuberculosis among males and a similar picture was also observed in the sample survey showing 4.5% males (out of 5.82%) in experimental and 30% males (out of 42.85%) in control group. Males are more likely to default the treatment increasing the morbidity and mortality among them. This affects the family and in particular females are sufferers. Tuberculosis kills more women in reproductive age group than all cases of maternal mortality combined and it may create more orphans than any other infectious disease. In India, women who suffer from tuberculosis face constraints as they tend to neglect their illness till they are too sick and they depend on others to get necessary medical attention.

Overcrowding, unsanitary habit, malnutrition, poor living condition and air pollution by the industries are the possible determinants for the causation of the disease.

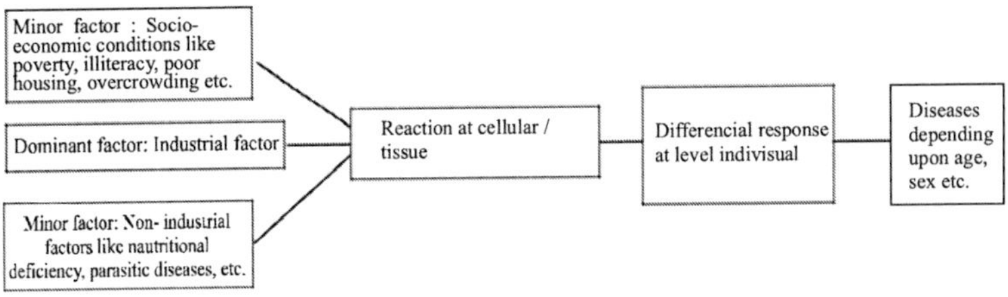

Figure 3: Interaction between factors and diseases

In control group, the occurrences of diseases were less as evidenced by questionnaire survey, clinical study and data from operational health care units. As the villages in this group were 20-25 Kms away from industry, chances of industrial impact were remote. Again the river Burhabalanga separates (Fig. 3) the control group from the experimental group by taking its course adjacent to the villages in control group. The existence of the river nullifies the industrial impact by the process of dilution.

However, there might be some other socio- environmental pathology operating among the experimental group whose health hazards arise not only from intrinsic disorder of bodily structure and function but also from environmental hazards related to industrialization and exposure to toxic substances and the evidence of such were discussed earlier.

Conclusions

By reviewing the results of survey it was concluded that industry has a significant role (Industry Factor) for the causation of disease which is supported by statistical analysis. Negative correlation establishes that more the distance of village from industry, less the impact of the industry factor. From statistical analysis it emerges that the impact of the industry is minimal on people residing away from industries and marked on those who are nearer to the industries. Although correlation can not be used to invoke causation, it may be said that causation implies correlation. It is reasonable to conclude industry factor having a "one-to-one" relationship as causation. However, different diseases existing predominately in the industrial villages might be the out come of interaction of some other minor factors; but industry factor remains a dominant one (Fig. 2). In addition, high prevalence of diseases in industrial villages assumes the increasing levels of exposure to the risk factor (Industrial Factor) concluding a biological gradient or dose-response relationship determining the strength of association. The result of clinical study, analysis of health data from the dispensary of the industries and from the health care units reveal that there is consistency in the result to establish a 'causal' association of industrial factor.

Considering the above conclusion (causation-strength of association- consistency), a health management model in the area of health management is the need of the hour comprising health risk awareness, health risk assessment and health risk mitigation and addressing to all the sectors like industry, gov-ernmental machinery and non governmental organization working in the environmental issues and population at risk.

Acknowledgments

The authors are thankful to Head, Dept of Zoology, Dr. Babasaheb Ambedkar Marathwada University, Aurangabad-431 004 (M.S.) India for providing laboratory and library Facilities. We are grateful to UGC for providing financial assistance as SRF during the course of study.

References

Anya, K. and Kay, T. 1994. Environmental and public health issues. In: Encyclopedia of Occupational Health and Safety, 4th Edn. ILO, Geneva.

Banarjee, B. P. 1985. Health and political system. Ind. J. Indust. Med.

28: 132-135.

Banarjee, K. and Banarjee, R. 2009. Occupational and industrial skin diseases. In: Valia, R.G. (Ed.), IADVL Text Book and Atlas of Dermatology, Vol. I, 2nd Edn. Bhalani Publishing House, Mumbai. pp. 498-509.

Brahma, S., Mohanty, J. K. and Das, M. 2003. Pollution in shallow aquifers in and around industrial estates near Balasore, Orissa. In: Assessment and Management of Water Resources. Society of Geoscientist and Allied Technologists, Bhubaneswar. pp. 299-310.

Chakravarty, M. K. 1967. Physiological response to thermal stress in a glass bangle factory. Ind. J. Indust. Med. 13: 131.

Clark, D. 1992. Occupational Diseases among industrial workers. Am. J. Ind. Med. 22: 395-409.

Dash, M. C. 2004. Ecology, Chemistry and Management of Environmental Pollution. 1st Edn. Macmillan India Ltd., New Delhi.

Donham, K. J. 1982. Acute toxic exposure to gases from liquid manure. J. Occup. Med. 24: 124-145.

Jarup, L. 1996. Epidemiology and Public Health, Imperial College, London, UK. 21: 289-298.

Marwari, R., Kala, M. and Khan T. I. 2009. Heavy metal analysis in Lycopersicon Esculentum grown in Sanganer area. The Ecoscan. 3 (1&2) : 69-74.

Phadtare, J. M., Ramraje, N. N., Khan, I. and Garde, S. 2005. Asthma-Principles of diagnosis and management. In: Shaikh, W. A. (Ed.), Allergy and Asthma: A Tropical View, IJCP Publications Pvt. Ltd., New Delhi.

Philipson, M. 1984. Occupational diseases among steel industry workers. In: Døssing, M. (Ed.), Occupational Liver Injury. pp. 102.

Ray, P. K. and Khangarot, B. S. 1989. The deadly metals. Science Reporter. 12: 560-561.

21

Concept of Green Mining

Introduction

In The Rio, Brazil conference 2012, 191 countries came together to debate and prepare a consensus document- 'The Future we want'. The Buzz word was Green Economy linking to Sustainable Development, and which had link to the Brundtland 987 report 'Our Common Future' & 1992-earth summit's buzz word of sustainable Development (Figure 1). UNEP defines it as 'Improved human well-being and social equity, while significantly reducing environmental risks & ecological scarcity by developing renewable energy technologies, energy efficient buildings and equipments, low-carbon public transport systems, clean energy vehicles, proper waste management with recycling facilities, green mining etc.

In this context Green mining concept emerged because mining is generally involves habitat destruction, rehabilitation of ecosystem people, resource depletion and environmental pollution. The environment is an 'enabler' of economic growth & societal well-being. Hence the environment is to be protected.

PILLARE OF SUSTAINABLE DEVELOPMENT

Hard Facts of Mining

India produces 89 minerals including four fuel minerals, 52 non-metallic minerals, 11 metallic minerals and 22 minor minerals. India stood as world's largest producer of mica blocks & mica splitting and second rank in Chromite production, third in coal & lignite, and barites, fifth in iron ore, sixth in bauxite and manganese ore, eleventh in aluminum and twelfth in crude steel in the World (IBM, 1999). With its contribution of about 20 percent in exports and >20 percent in imports, the mineral sector is an important component of India's foreign trade. Mining and mineral based industries are vital to economic growth of nations.

Foreign investments flow to this sector. The importance of the sector in national economy is significant.

However, Impact on Environment of mining activities can leave deep scars on the earth's surface and damage the ecological and social systems. Naturally, these damages inhibit long term prosperity and are the negative factors of sustainable development.

Impacts of Mining on Environment

Visible impacts on Land, water bodies, ecology of the site are seen. Impacts also occur in the atmosphere and on the society. Removal of vegetation, affects fauna. Pollution of water bodies affects aquatic ecology. High fugitive emissions and release of other gaseous emissions-retard growth of vegetation and adversely affects human health. Noise pollution due to blasting and subsequent Vibration- drives away wild animals. (Figures 2- 5).

Fig. 2

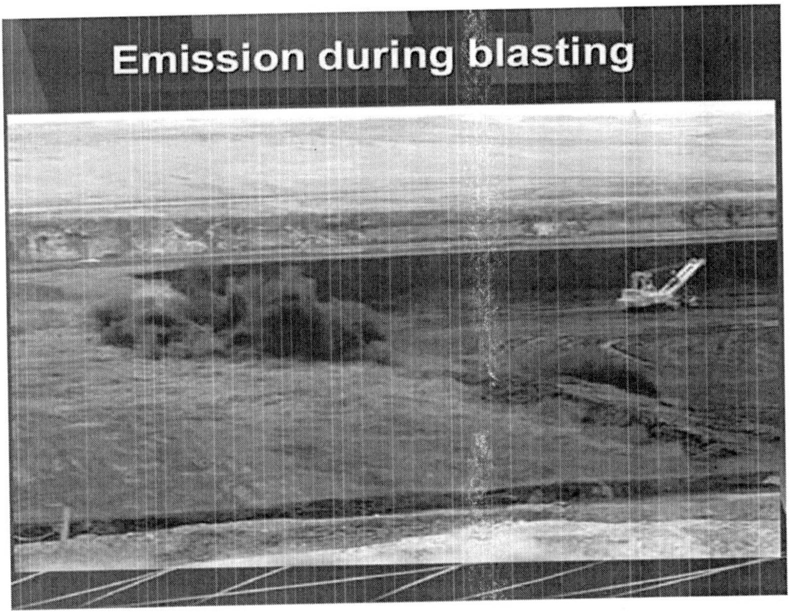

Fig. 3

Fig. 4

Fig. 5

Policy initiatives taken in India to meet the challenges:

(i) Disallowing mining in forests to the extent practicable/ providing minimum area required for mining in forest area.

(ii) Fixing stringent conditions of compensatory aforestation in immediate vicinity to sustain forest cover.

(iii) Framing stronger and scientific mine closure regulations

(iv) Policy to support in the form of environmental laws under Water Act-1974, Air Act-1981 and E.P.Act-1986 for the protection, maintenance, and rehabilitation of natural habitats were taken. Under EP-Act, EIA rules were framed. The EIA process operates in India since September 1994 in accordance with the resolution taken in the Earth Summit in 1992.This process be reinvented and integrated with the mining approval process to get results leading to green mining. *Since many green technologies for the plantations, energy use, and abatement of pollution have been developed, stress is given to utilize these technologies so that green mining is achieved.*

The Environment Impact Assessment needs stringent compliance, transparency, monitoring and accountability. The problem of mine closure after mining needs to be addressed with effective regulatory monitoring and with penal provisions so that green mining concept is realized. Independent and impartial preparation of Environmental Impact Assessment (EIA) reports is yet to be introduced.

Green Mining Practices

Green Mining Mission 2012- 2016 practices involve the following aspects:

(i) Shutting down illegal and unregulated mines, (ii) Choosing environmental friendly general mining processes such as use of Rapper Dozer instead of blasting in surface mining, use of Surface miner, etc, (iii) Implementing recently discovered green mining technologies for over-burden dump slope management using coir matting and specified herbaceous species such as Citronella sp./Reeds, plantation techniques using multi species (CPCB-2000), using horticulture practices, developing nurseries, water management, etc., (iv) Use of ecofriendly processes for Cleaning up the mining sites of shut-down mines, (v) Research and Development of Green Mining Technology.

Case Studies:

The rare —earth mine in the Clark Mountain Range in California, USA called Mountain Pass mine was producing some of the world's supply of rare earth metals. It was owned by Chevron. Mountain Pass mine in the United States was operating at full capacity in the 1990s, and produced as much 1.224 million gallons of salty wastewater per day, containing radioactive thorium and uranium" (Molycorp, 2011). This waste leaked into the desert, harming the local environment and costing Molycorp hundreds of millions of dollars. The mine was shut down in 2002.

Molycorp reopened the mine at full capacity of 20, 000 MT a year in 2013. Some major changes in their environmental mining policy since closing in 2002 include: (i) Changes in tailings storage. Paste-tailings are now dried out and laid flat in a facility, which makes them less volatile and more compact, This reduces the supply of fresh water, since the water from tailings can be recycled, (ii) Chemical reagents in tailings are recovered and recycled along with the water, (iii) New technologies which use the excess cerium in stockpiles are used and cerium's magnetic properties remove phosphorus from water ("Molycorp annual report, " 2011). iv). Using waste heat from mining to generate steam and power, and thus decreased the carbon footprint. These changes also reduced Molycorp's production cost by saving on water, chemical reagents, heat, power and total area used for stockpiling and waste piles. As such, these policies are both environmentally friendly and economically feasible. Molycorp increased funding to $8.3 million for research and development. This research funding helped to develop some of the new technology listed above (Molycorp Annual Report, 2012). Green Mining Mission 2012-2016 proposes building from these precedents and adapting them to fit the evolving needs into the future to make mining of rare earths more environmentally sustainable and financially viable. *(Google.com search: green mining*

web.mit.edu/12.000/www/m2016/final, website/.../greenmining.html, www.nrcan.gc.ca/mining-materials/green-mining/8178, www.canadiangeographic.ca/themes/environmental.../green-mining.asp, en.gtk.fi/mineral resources/greenmining.html, www.mining-technology.com/.../feature environment-friendly-mining-tre... Jan 30, 2014, Dust suppression solutions. (2003). Retrieved from http://www.wetearth.com.au/dust-control-suppressionGreen mining initiative.

(2012,). Retrieved from http://www.nrcan.gc.ca/minerals-metals/technology/4473)

The recent technologies for extraction of rare earths include:

(a) **Mining from tailings:** Sometimes amounts of the metals being mined end up in the wastes. Reprocessing these wastes can result in more usable material,

(b) **Liquid membrane emulsion technology:** A separation technique that can be used to extract usable metal from highly toxic or acidic waste water produced by a mine. This technology can even deal with dilute solutions,

(c) **Dust suppression techniques:** By pre-wetting the areas to be blasted with high volume sprinklers is an option to reduce dust. Once the dust is in the air, the only way to remove it is through mist cannons. Open pit mining techniques can reduce dust through by purchasing high volume sprinklers and Wet Earth Fog Cannons.

(d) **Sulphuric acid leaching extraction process:** Currently, bastnaesite is generally processed in a hydrochloric acid leaching process. This process leaves fluorine and thorium contaminated wastes and it creates cerium of low purity. However, a sulphuric acid leaching extraction process can produce high purity cerium while also allowing thorium to be separated for use and better containment (Schuler, Buchert, Liu, Dietrich & Merz, 2011),

(e) **Impermeable tailings storage:** Tailings are often stored in some form of tailings pond. Tailings will be required to be stored in impermeable tailing ponds to avoid ground and surface water pollution. Alternatively, the tailings can be dehydrated into a paste, as done by Molycorp (Molycorp Annual Report, 2011),

(f) **Choice of ventilation and diesel engines:** 10 percent of mining costs come from electricity consumption. Better ventilation systems and diesel engines could be purchased by mines to maximize efficiency. The Canadian Mining Initiative has created a list of approved diesel engines (Green Mining Initiative, 2012).

Mission –Green Mining 2012- 2016 suggests that an international non-governmental environmental group concerned with mine clean-up and implementation of green mining technology will be started and funded by governments, private donations, proceeds from reclaimed materials, and mining companies. Mining companies will have interest in this group because it will provide a rating for mines based on how minimal their environmental impacts are. The governments shall have to grant tax breaks to mines with especially high ratings. This rating should be accepted in India by the mining companies.

Towards Green Mining in India:

Chromite Ore Mining in Sukinda Valley in Odisha started in 1949 when there was no environmental laws to abate environmental pollution, no treatment facility, no plantation techniques etc etc.The valley was largely abused from environmental protection point of view (Tewari and Naik.2014)

Current environmental problems:

(1) Mine overburdens, and their management.

(2) Beneficiation of Ore-waste water in the form of tailings with Hexavalent chromium & overburden erosion causing Water pollution (Hexavalent chromium),

Earlier 22 mines were operating with a leasehold area of 583.021ha. At present 13 mines operate and the annual production is ~4 million tons. (World production.~460 million tons-Kazakhstan 210 MT, South Africa-200 MT, USA 0.62MT).

State Pollution Control Board (SPCB), Odisha played the key role-since late 1990s- directions were issued & at present all mines have Effluent Treatment Plant, & chrome ore beneficiation is done scientifically. Overburden dump management with plantations, slope stabilization with coir matting, etc is strictly practiced. Forests are regenerating due to protection. SPCB gave direction to the mines in 1990s-2000 to install ETP and treatment plants and this has been complied with. The technology involved used Ferrous Sulphate to precipitate Chromium Oxide and Sodium Hydroxide was used for pH corrections (Figure 6).

SPCB, Odisha appointed IIT, Kharagpur to review the existing technology and suggest improvement. Their suggestive model of treatment of mine drainage water has been implemented and given in Figure 7.

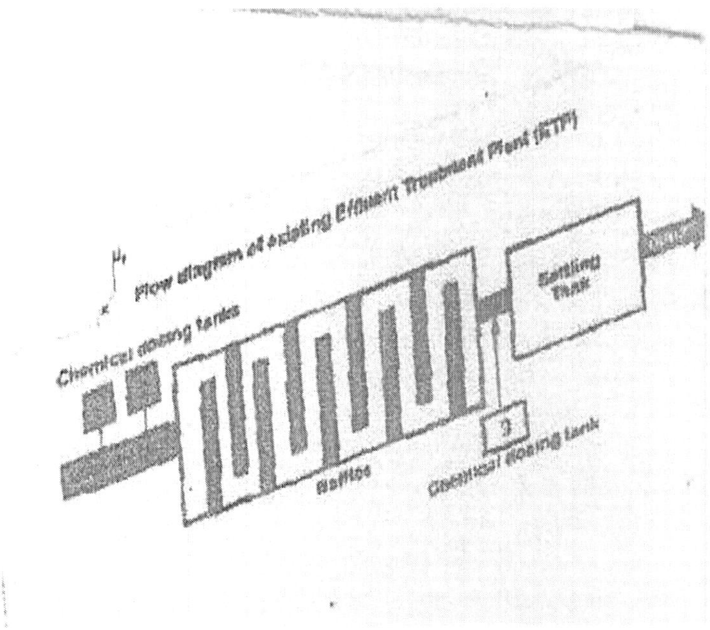

Fig. 6. The flow diagram of ETP (old method)

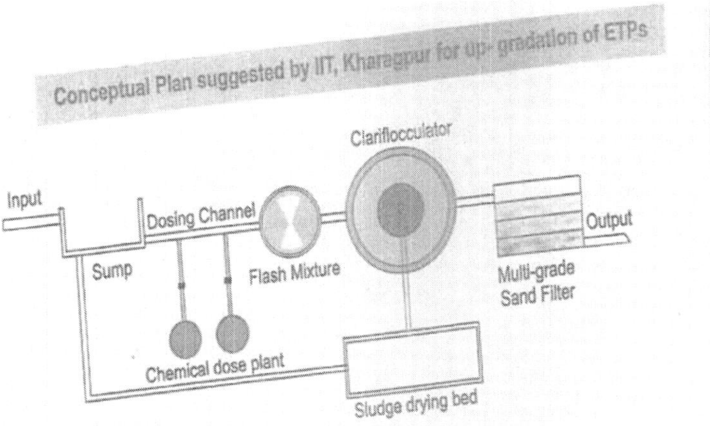

Fig 7. IIT, Kharagpur model for Chromite mine drainage water Treatment (SPCB, 2014)

Besides the mines were directed to go for plantations as per rule and were advised to use coir matting and herbaceous plantations in overburden dumps as shown in the figure-8.The mines have complied to the direction. The periodical monitoring indicates that the Hexavalent chromium level in Damsalla nalla and Brahmani River is much below the prescribed standard.

Indian Case Study-2: Talabira coal mine phase 1.

It is the first Blast free coal mine in India.Ripping Technology adopted for the first time in ripping of sand stone. The mine was working below the Highest Flood level of Hirakud reservoir till 2014, when the mine was closed. No electrically operated machinery used in the mine.No accident has occurred since inception (~20 years)

Green technology was used for overburden slope plantations as shown in the figures 8-10.

Overburden dump plantation (Figure 8 – 10) is good. The Talabira mine phase-I has been awarded all India environment protection prizes and environmental excellence prize from SPCB, Odisha. (photo from Er.Sushant Guru)

Fig. 8

Fig. 9. Stabilised OB Dump of Talabira phase-I coal mine, Odisha

Fig. 10. OB Dump plantation (From Er Sushant Guru)

Conclusion

Unfortunately, Green mining was not widely practiced in India. Although the EIA process has important components of green mining but it was not practiced with full commitment. At present the thrust is green mining practicing with commitment. It is high time for the industry proponents, the law makers, regulatory agencies and the intelligentsia to seriously consider and debate over the vital issues and prefers a sustainable development model of mineral development with green mining for long term benefit to the nation.

References

http://www.inlandsocal.com/business/content/manufacturing/stories/PE_News_Local_S_molycorp 09.44b8a72.html.

Google.com search: green mining

web.mit.edu/12.000/www/m2016/final website/.../greenmining.html

www.nrcan.gc.ca/mining-materials/green-mining/8178

www.canadiangeographic.ca/themes/environmental.../green-mining.asp

en.gtk.fi/mineral resources/greenmining.html

www.mining-technology.com/.../feature environment-friendly-mining-tre... Jan 30, 2014

Dust suppression solutions. (2003). Retrieved from http://www.wetearth.com.au/dust-control-suppression

Green mining initiative. (2012,). Retrieved from http://www.nrcan.gc.ca/minerals-metals/technology/4473

Hartman, H., & Mutmansky, J. (2002). Introductory mining and engineering. John Wiley & Sons Inc.

Hogan, L. (2004, 01). Research and development in exploration and mining. Retrieved from http://adl.brs.gov.au/data/warehouse/pe_abarebrs99001016/PC12624.pdf

Molycorp annual report 2011. (2011). Retrieved from http://www.molycorp.com/wp-content/uploads/2011-annual-report.pdf

Molycorp innovations. (2012). Retrieved from http://www.molycorp.com/technology/molycorp-innovations/

Paul, J., & Campbell, G. (2011). Investigating rare earth element mine development in epa region 8 and potential environmental impacts. Retrieved from http://www.epa.gov/region8/mining/ReportOnRareEarthElements.pdf

Reuse and re-process (r2) technologies. (n.d.). Retrieved from http://www.itrcweb.org/miningwaste-guidance/to_reuse.htm

Schuler, D., Buchert, M., Liu, R., Dietrich, S., & Merz, C. (2011). Study on rare earths and their recycling. Retrieved from http://reinhardbuetikofer.eu/wp-content/uploads/2011/01/Rare-earths-study_Oeko-Institut_Jan-2011.pdf

Topf, A. (2011). Company proposes to mine Arizona copper deposit using in-situ method. Retrieved from http://www.metalaugmentor.com/reviews/company-proposes-to-mine-arizona-copper-deposit-using-in-situ-method/

Ulmer-Scholle, D. (2012, June 27). Uranium- how is it mined?. Retrieved from http://geoinfo.nmt.edu/resources/uranium/mining.html

Voigt, K. (08, 08 2012). China cuts mines vital to tech industry. Retrieved from http://edition.cnn.com/2012/08/08/business/china-rare-earth/index.html.

Guidelines for Developing Green Belts.2000.Central Pollution Control Board, New Delhi, 205pages.

Tewari, L.K. and C.R.Naik.2014 Chromite Mining and State of Environment in Sukinda Valley of Odisha, Odisha Pollution Control Board, 406 pages

SPCB, Odisha, 2014 (personal communication)

Part VI : What is Scientific Temper?

22

Fostering Scientific Temper

Introduction

We are living in an era of science and technological development. Science has radically changed man's material environment. Science and technology have fostered a new intellectual temper known as Scientific Temper. Every Nation desires to be a 'knowledge society' based on science & technology.

The term science is derived from the Latin verb 'scire', meaning 'To whom' & Latin word 'scientia' meaning 'knowledge'. Science is a systemized body of knowledge- As per dictionary meaning science is accumulated & systemized learning to natural phenomena & includes observation, methods of collecting information, analyzing the data qualitatively and quantitatively and drawing conclusion based on analysis of observed data.

Beauty of Nature:

How to understand and appreciate the intellectual beauty of the universe?

We may consider the Matter-Energy Matrix and Look into the Diversity-

Diversity & complexity of matter, diversity of living matter, living thinking matter, Living thinking matter with intellect. We may appreciate the forces/energy that pervades the universe & operates-Gravitation, Electromagnetic, Intra & Intermolecular forces. Energy-One form can be transferred to another- not created nor destroyed.

Is there unity among these diversities?

Human Endeavour

There is diversity of human endeavor to appreciate the beauty, diversity, unity in the diversity, the truth of creation of this wonderful universe/ Multiverse. These endeavors are:

(i) spiritualism, (ii) Art & Sculpture, (iii) literature, (iv) music, (v) dance, (vi) Philosophy, (vii) Science- Scientific studies appreciate the creation and tries to understand and appreciate the functioning of the universe - matter & energy matrix and science has predictable quality as it based on quantification.

Is it not challenging? Is it not worth pursuing?

Science deals with quantification and based on trend of occurrence of phenomena & quantification, it can predict things to come. Its inherent quality creates human curiosity, thus lead to (i) observation, (ii) hypothesis, -Theory (iii) measurement, (iv) experimentations, (v) analysis and (vi) conclusion & communication. Science develops capacity of reasoning, thinking and questioning.

Scientific Temper

Jawaharlal Nehru was one of the first people to use the term scientific temper and advocated the promotion of scientific temper:

"[What is needed] is the scientific approach, the adventurous and yet critical temper of science, the search for truth and new knowledge, the refusal to accept anything without testing and trial, the capacity to change previous conclusions in the face of new evidence, the reliance on observed fact and not on pre-conceived theory, the hard discipline of the mind—all this is necessary, not merely for the application of science but for life itself and the solution of its many problems." — Jawaharlal Nehru (1946) *The Discovery of India*, p. 512.

Scientific temper is a way of life – of an individual and social process of thinking and acting - which uses a scientific method, narrated in the previous in the previous chapter.

The Story of Samrat Ajatsatru and his Guru 'Gargya Muni -Guru wanted to teach him about God. Ajatsatru said I shall gift you 1000 cows if I shall learn anything new, which I do not know ?

Guru Questions -Ajatsatru Answers

1. Soul of sun is god–Sun gives us light and heat,
2. Soul of moon is god-Moon brings peaceful attitude and desires to drink wine,
3. Soul of air and fire is god–Air is comparable to my army, who provides me security and fire is comparable to my desire to have an empire.
4. Soul of Sound and shadow is god–Sound is life and shadow is death
5. Soul of man and living Organisms is god–Body nurtures the soul.

Was not this the scientific temper in Ajatsatru?

The scientific temper represents a spirit of query and as such involves the process of logical reasoning. It tries to find out the cause and the rational justification of an event objectively and is free from all prejudices and make-believe. It would attempt to explain the cause-effect relationship between two or more events. Thus a man with scientific attitude would not like to take for granted and believe in irrational beliefs and superstitions. Such beliefs differ from society to society according to their culture, ethos and heritage and the attitudes of the individuals. What is auspicious to one group may be inauspicious to another. Thus, being irrational, they lack universal applicability.

Scientific knowledge, which is true and objective, must possess the qualities of universality and verifiability.

Nehru wrote that scientific temper goes beyond the domain in which science is normally done, and deals also with the consideration of ultimate purposes, beauty, goodness, and truth.

He also said that it is the opposite of the method of religion, which generally relies on emotion and is (mis ?) applied "to everything in life, even to those things which are capable of intellectual inquiry and observation." While religion may tend to close the mind and produce "intolerance, credulity and superstition, emotionalism and irrationalism", and "a temper of a dependent, unfree person", — Scientific temper "is the temper of a free man"- open mindedness.

Nehru indicated that the scientific temper goes beyond objectivity and fosters creativity and progress.

He envisioned that the spread of scientific temper would be accompanied by "the exciting adventure of fresh and never ceasing discoveries, of new panoramas opening out and new ways of living, adding to [life's] fullness and ever making it richer and more complete." The genesis and development of the idea of scientific temper is connected to ideas expressed earlier by Charles Darwin when he said, "freedom of thought will best be promoted by that gradual enlightenment of human understanding which follows the progress of science, " One of the fundamental duties of Indian citizens, according to the Constitution of India is "To develop scientific temper".

One great merit of the scientific method is that it adopts a self corrective process. In fact, all sciences progress by disproving, revising and sometimes rejecting the earlier ideas, laws, theories and beliefs in order to gain more and more definite and comprehensive explanation of things. This helps in cultivating such essential qualities of the human mind like sustained effort, detachment, tolerance, objectivity and creativity, which are essential equipments for solving the personal, social and international problems of men. A man with scientific temperament is expected to be always open minded and as such he is never under the illusion that science and technology can provide a panacea to human ills.

Science has given us vast knowledge, which has brought us immense power. But it is man who uses that knowledge and man who wields that power. Every nation desires to be Knowledge society, which brings societal transformation, generates wealth and brings happiness & comfortable living & thinking of cosmic imperative.Science is neither good nor evil *per se.* But man is a moral being, and on his choice depends the course of science and the future destiny of humanity.

The crisis we face is a moral one, for it is the outcome of the conflict within man himself. Bertrand Russell has rightly said: We are in the middle of a race between human skills as to means and human folly as to ends....Unless men increase in wisdom as much as in knowledge, increase of knowledge will be increase in sorrow and therefore the need of the hour is for men of wisdom. Mere knowledge without wisdom to guide us in the utilization of that valuable knowledge will make for greater sorrow. Are we different from other Primates?

Comparison of present day primates and man:

Comparison of Present Day Apes & Man

•	Gibbon	Orangutan	Chimpanzee	Gorilla	Man
1. No.of Species	4	1	1	1	1
2. Subspecies/Races	15	2	-	2	30
3. Body weight Average (LB)	11-21	110-165	125-150	150-450	110-160
4. Height Average (Feet)	2.3 (M), 3 (F),	4.8(M) 4.3(F),	5.0(M) ,4.3(F),	6.0(M) 4.1-5.4(F) ,	5.4- 5.8M (Bipedal)
5. Cranial capacity (Cubic inch)	5.95-7.6	23.5-30	23.1-27	29-38.3	61-115 (abundant synapses)
6. Sexual maturity (Years)	5-8	10-12	7-12	7-10	12-17
7. Gestation (days)	200-212	233	202-260	268	280
8. Longevity(years)	30	30	35	30	60-85
9.Courtship & foreplay	√ minimum	√ minimal	√ minimum	√ medium	√ Extensive
10.Sex (copulation)	From Back side	Back	Backside	Back	Front- (interaction-deep)

Think the impact of Science & Technology in Transport, (Aeroplane) Communication (Microprocessor-TV etc), Medicine (Antibiotics), Gene Technology (Human Genome deciphering & linking to diseases, Finger Printing etc), Nuclear Fusion (Source of enormous energy), Medical Technology (Scanning, X-rays, Magnetic Resonance Image etc) Biotechnology, GM crops, Pollution Control etc. In 1900-1950, the emphasis was on subatomic particles, understanding the structure & organization of the universe, galaxies, chemical evolution, the physical basis of heredity, human genome decoding, understanding the order in ecosystems, etc.

By 2000, many types of subatomic particles identified, human genome decoded, estimated of galaxies increased from one, our own Milky Way to many billion.

The present human civilisation and our day to day living is science-Technology centered. Human brain's quest for acquisition of knowledge, inquisitiveness, questioning and analyzing ability is the key to acquisition of new knowledge & development of science & technology. Twentieth century witnessed spectacular advances in physical science, biological & bio-medical science including Genomic Science, & technology including understanding the functioning of natural ecosystems and on environmental technology. Every nation desires to be a knowledge society. Why one, fully aware of one's inherent & often insurmountable limitations, nevertheless devotes oneself to scholarship & to a life of constant endeavour, many failures & few successes? The answer-T.S.Eliot has given in his 'Confidential Clerk': that a man should have a consuming passion/ To do something for which he lacks the capacity.

Knowledge Leader

Our great country was the leader of the world knowledge. Great physicians and surgeons like Jewaka, Charaka and Vaghbhata, Samanta Chandra Sekhar belonged to this country. The celebrated Indian astronomers and mathematicians like Aryabhat, Brahmagupta, Bhaskara, and Varahamihira, worshipped the Goddess of knowledge. Scholars like Parini, (Palini), Patanjali, Kanad and Shankara imparted knowledge to students and scholars from all the corners of the world who came to India in search of knowledge and excellence. We all are proud to be part of this knowledge heritage. Aryabhat invented Zero digit in Number System,

Aryabhat was also the first to explain spherical shape, size, diameter, rotation and correct speed of Earth in 499 AD. Bhaskaracharya (5th century) rightly calculated the time taken by Earth to orbit the Sun, hundred years before the astronomer, Smart. Bhaskaracharya calculated that the earth takes 365.258756484 days to orbit the sun. Ayurved is the earliest school of medicine known to mankind and was consolidated by Charaka 2500 years ago. It is claimed that the art of Navigation was born in the river Sindh some 6000 years ago. The word 'Navigation' is derived from the Sanskrit word NAVGATH.

Knowing is not enough, we must apply, willing is not enough, we must do it. Disciplines as thrust areas in 21st century:

(i) Material Science, (Nanoscience, Nanotechnology, Artifical human organs etc)

(ii) Modern Agriculture, Agricultural Zoology, Soil Biology & Biodiversity

(iii) Alternate source of energy, (Solar, Wind, Tide, Biomass gasification, etc.)

(iv) Climate Change-Carbon sequestration & forest conservation, Technology linked to C-sequestration)

(v) Information Technology, Astrophysics & Space technology

(vi) Biomedical sciences

(vii) Molecular biology, Biotechnology, Bioinformatics, Stem cell research etc

(viii) Ecosystem analysis, Life cycle Analysis, Biodiversity Conservation, Movement of materials & pollutants in ecosystems,

(ix) Environmental Technology to abate environmental pollution & degradation and so on...

(x) Management Science, Environmental Management & sustainable Development,

(xi) Cottage industry, & Metal works, & application of new technology to these industries,

(xii) Preserving Traditional knowledge and applying for human welfare.

Science & Technology Knowledge develop Skill and Competence. The curriculum and the pedagogy lack diversity and flexibility respectively, thus creating a big gap in the demand supply chain. How to fill up these gaps?

It is only through increasing the know-how of the students passing from an institute. Scientific knowledge is the most dynamic concept in the changing market scenario. The main issues for the school, Colleges & Universities are to develop quality students who can be the potential scientists & technologists/managers of the society to look after its welfare. In view of this quality teachers are required.

Scientific knowledge develops a sense of curiosity & questioning in the mind of pupil. It is desirable to train them to acquire the capability of reasoning and thinking, and to create open-mindedness among the students- free from dogmas, superstition, blind faith etc. Bring the students out of superstition by reasoning and experimentation, and It may be desirable to use problem solving technique so that student develops the capacity of application. It may be profitable to inculcate the values/Ethos on which Indian culture stands along with scientific temper.

Science Advances

The first half of 20th century witnessed new invention/discovery in Physical Sciences- X-rays, understanding of the structure of atom, chemical bonding, quantum mechanics, theory of relativity, the Earth cycles, etc.The later part of first half witnessed new understanding of Structure of cell & chromosomes, discovery of antibiotics. Mendel laid the foundation of genetics in 19th century, and Morgan, Muller & others carried forward- The gene theory, and elucidation of protein structure, & DNA model (Watson & Crick). Systems approach to understand micro, macro & mega natural and biological systems.

The second half of 20th century witnessed invention of silicon chips, & communication technology, Computer technology, Internet, further understanding the structure of atom, space technology & further understanding the structure of the Universe, Biotechnology, & Bioinformatics, Genetic Finger Printing, & Transgenic- The Genomic Era & Beyond, Human genome enumeration, & linking genome to diseases and human evolution, Scanning technology, & Advancement in Neuroscience & Health Care (Magnetic Resonance Imaging, Positron Emission Tomography, CAT (Computerized Axial Tomography etc, Environmental Technology to abate pollution, Climate change, to increase food production and providing food and protein security, Material science- Nanoscience & nano-technology

International Human Genome project (HGP) deciphered the book on human life in 2001 by translating about 3 billion strands of DNA & they found that the human genome/genetic code is 99.9% identical in all people. There are about 25000 genes which code more than 20000 numbers of proteins in humans. The 0.1% difference in DNA sequencing is responsible for our differences to susceptibility to diseases, eye colour, hair colour, Athleticism etc. This finding supports the theory that humans had a common ancestor with the earliest humans who lived in Africa (possibly to a single human female-Mitochondrial Eve).

Nanostructure materials are being used for production of more efficient and inexpensive solar cells. Solar derived energy can be used to power the electrolysis of water to produce hydrogen, a clean energy.

Hydrogen being the smallest element, likely to escape from storage tanks and pipes more easily than the conventional fuels. Very strong materials are needed to store hydrogen at very low temperature and high pressure. Novel nonomaterial such as carbon nanotubes have to store upto 70% of hydrogen by weight, an amount 20 times larger than the compounds that is used currently.

Carbon nanotubes are 100 times stronger than steel at 1/6th of the weight.Thus a 100- pound container made up nanotubes can store as much hydrogen as could 600-pound steel container and the walls would be 100 times as strong. The other thrust areas are: Molecular Biology:- A Revolutionary Science Biotechnology, & Bioinformatics, Stem cell research, Use of Transgenic for 2nd green revolution, Understanding human genome ——disease relationship-Biomedical Research, Ecosystem Analysis & Land use Management:- A new tool for 21st century- Understanding of the ecosystem functioning- Homeostatic mechanisms & Stability, . Bio-resource conservation & management, Biodiversity, Adaptation biology & Climate change.

These thrust area disciplines depend upon strong background in Basic Sciences.

Traditional breeding led us in to the trap of narrowing down genomes

(Donor plant with desired genes → x Commercial plant → new plant variety with many transferred genes: traditional practice) – may not assure pest resistance, assures increased yield, Post Genomic period-Biotechnology could bring back at least that part of genetic diversity which enhances pest resistance and also greater yield (Donor with desired gene → commercial plant + with the desired gene → A single gene is transferred to the improved commercial plant.).

Biosafety & Toxicological aspects are needed for immediate review.

GM crops after lab and field trial ready for release in India

Rice-	Bayer Bioscience, Avesthagen, Mahyco & Metahelix Life Science
Mustard	Delhi University
Corn-	Monsanto, Pioneer overseas & Dow Agro- sciences
Sorghum-	National Research Centre for Sorghum

Groundnut- ICRISAT

Cauliflower- Sungro Seeds & Nunhems India

Tomato- Avesthagen, IARI & Mahyco

Potato- Central Potato Research Institute

Brinjal- Mahyco, Bejo Sheetal Seeds, Sungro Seeds

Cabbage- Nunhems

Okra- Mahyco

There is now research cooperation between private sector & Institutions in India.

Research on transgenic plants and animals for sure will be thrust areas for future. Transgenic invasion & Bio-safety, & Antioxidant research are important fields. Intellectual Property Rights, Patenting & Ethical Issues are important aspects for consideration. Political- will is very important to bring change & development. Are our universities/ Institutes prepared to face this challenge?

References

Encyclopedia of Life Support System, EOLSS Publishers, Oxford, 2002.

Nuffield Bioethics, UK GM Crops Discussion Paper 2004.

WWF Report 2009,

India Today- 2 November 2009

Publications of ABLE, 2010, 2011

The Week, 25 November 2009

Down to Earth, 2011 issues

From Wikipedia, the free encyclopedia:Darwin, Charles. "Darwin Correspondence Project". *Darwin, C. R. to Aveling, E. B. 13 Oct 1880.* Retrieved 29 August 2013. "Scientific temper and the argumentative Indian". Chennai, India: The Hindu. 2005-09-22. Retrieved 2007-09-25.

Basu, Durga Das (1993). *Introduction to the Constitution of India* (15th ed.). New Delhi: Prentice Hall of India. p. 131. ISBN 81-203-0839-5.

23

What is wash?

Introduction

WASH stands for "Water, Sanitation and Hygiene" - several interrelated public health issues that are of particular interest to international development programs and is a key public health issue, especially in developing countries like India . The concept of WASH involves together water, sanitation, and hygiene because the impact of deficiencies in each area overlaps strongly. In view of this, the thrust is to address these deficiencies together to achieve a strong positive impact on public health. Affordable access to WASH is must for sustainable development.

The UN's Millennium Development Goals of 2000 AD included improvement of WASH services to target population without sustainable access to safe drinking water and basic sanitation. This is part of the Sustainable Development Goals where Target 6 includes "Ensure availability and sustainable management of water and sanitation for all". Several international development agencies have identified WASH as an area with significant potential to improve health, life expectancy, student learning, gender equality, and other important issues of international development.

WASH is also linked to the poverty. After years of anticipation the Millennium goals are with us that 15 years to end poverty and 15 years to achieve something extraordinary that the first generation where everyone in the world has clean water and to be the last generation to witness children dying of diarrhoea and water borne diseases. Lack of sanitation contributes to approximately 700, 000 child deaths every year due to diarrhea, and chronic diarrhea can have a negative effect on child development (both physical and mental). In addition, lack of WASH facilities can prevent students from attending school, impose a burden on women, and diminish productivity. Access to safe water, adequate sanitation, and proper hygiene education, can reduce illness and death, and also impact poverty reduction and socio-economic development.

To meet the challenge of bringing water to 660 million and sanitation to 2.4 billion who don't have it now in the world, and to cater for the 1.2 billion new people who'll be sharing the planet with us by 2030, will require change and planning.. There's very little chance that we'll achieve the goals by 2030 if we do not bring the change at a very fast and plan for the future. Therefore the stress is to reduce the time counting down five years instead of 15 - five years to get going on meaningful change; so that in fifteen years time we really will get water and sanitation to everyone, everywhere, and forever. In view of this, the role of NGOs, donors, the private sector supporting and Governments is important. The United Nation's International Year of Sanitation held annually since 1993 and especially focused in 2008 has helped to increase attention for funding of sanitation in WASH programs of many donors. For example, the Bill and Melinda Gates Foundation, has increased their funding levels of sanitation projects since 2009, with a strong focus on reuse of excreta. The estimate is that 1.5 billion People work in water-related sectors (1).Women are hit the hardest when WASH (water, sanitation, and hygiene) infrastructure is inadequate. When the drinking water source is not on the premises, an adult female is the person typically collecting water (Table 1), and Plate 1.

Table 1. Persons (%) usually collecting drinking water from the away from home area (Bangladesh, 2012 & 2013- BBS & UNICEF, 2015).

Population Categories:	Adult Woman	Adult Man	Female Child	Male Child	
	88.8	5.4	3.7	0.9	1.1

Globally, it is estimated that women and children spend 125 million hours each day collecting water. Time is money and it adds up. This time could instead be spent generating income or doing other activities.

Plate 1 Girls going to long distances to collect drinking water

By alleviating this time consuming chore, we're giving women more hours back in their day and empowering them to use their time as they see fit.

There's a real opportunity to improve health and economic development through a comprehensive WASH strategy. According to data compiled by the World Health Organization's and UNICEF's Joint Monitoring Programme for Water Supply and Sanitation (JMP), the economic return on drinking water expenditure in South Asia, which includes Bangladesh, is the highest of any region in the world .

For every $1 invested in drinking water sources in S. Asia, an economic return of 2-4 times is generated (Table 2)

Table 2. Benefit-cost ratios of intervention to attain universal access of improved drinking water sources by regions, 2010, see the reference list)

Country/Region	Ratio
North Africa	2.4
Latin America	2.4
East Asia	1.6
South Asia	3.7
S.E.Asia	0.9
West Asia	2.3
Oceania	0.6
World Average	2.0

This means a $10 donation to North Africa, S.Asia including Bangladesh drinking water projects can not only provide a person with clean drinking water, but it can yield an economic return of over $24 to 37!

Health Issues: Neglected Tropical Diseases

Water, sanitation and hygiene interventions help to prevent many neglected tropical diseases (NTDs), such as trachoma, lymphatic filariasis, dengue, schistosomiasis, and helminthiasis. In August 2015 the World Health Organization (WHO) unveiled a global strategy and action plan to integrate WASH with other public health interventions in order to accelerate elimination of NTDs, especially elimination of trachoma and lymphatic filariasis as public health problems by 2020, and intensified control of dengue, schistosomiasis and soil-transmitted helminthiasis. The aim is to use synergies between WASH and NTD programmes. In developing countries, Healthcare can be associated with WASH and is in immediate need of reforms. It should be more accessible, affordable and transparent for the vast majority of the citizens.. What are the main issues in our healthcare system? Medicines and medical tests are getting expensive day by day. A nexus has developed between the hospitals / doctors and the diagnostic centers and path labs. This nexus is to be broken by sensitization and by law.

The strategic plan consists of four strategic objectives: (i) Improving awareness of benefits of joint WASH and NTD actions; (ii) monitoring WASH and NTD actions to track progress; (iii) strengthening evidence of how to deliver effective WASH interventions; and (iv) planning, delivering and evaluating WASH and NTD programmes with involvement of all stakeholders.

Sensitization:

Sensitizing Students in School and people living in urban slums, and people of the weaker section of the society is very important to get tangible results.

Case Study 1:

In a recent study by Hunter, Prüss-Ustün. 2016 on the question, 'Have We Substantially Underestimated the Impact of Improved Sanitation Coverage on Child Health? A Generalized Additive Model Panel Analysis of Global Data on Child Mortality and Malnutrition, published in PLoS ONE 11 (10) : e0164571. doi:10.1371/journal.pone.0164571, they have concluded that :

Data on population, improved sanitation and water coverage and all cause childhood mortality were available for almost all countries from 1991 onwards to 2014. Data on the two malnutrition variables (stunting and underweight) were available in only certain years which varied from one country to another. Data on diarrhoea mortality was available for the years 2000, 2005, 2012 and 2015. The data were analysed .In conclusion based on robust statistical analysis, their results from the data collected from World Bank- mostly Developing economies from Odisha, other parts of India, Bangladesh and South-Asian countries and elsewhere, would suggest that increasing sanitation coverage is one of the more important interventions for reducing neonatal and children mortality globally and would dwarf the impact of increased wealth and health expenditure. However, most of the health gains would be achieved by increasing community WASH coverage above about 80%.

Their results also suggest that increasing sanitation coverage would be a major contributor to reducing childhood stunting and underweight, though only after coverage exceeds about 50% and improved water supply coverage is probably more important. They also suggest that most prior studies have underestimated the impact of sanitation because they have focused on personal toilet use as a predictor rather than community coverage which we consider the more important. The analyses suggest that both universal access to improved sanitation and improved water is necessary for the biggest health benefit, but that gains are more likely to be seen with earlier focus on sanitation improvements. Future studies of the beneficial impact of sanitation on child health should use community coverage with improved sanitation as the primary prediction. There is scarcity of water in rural India, especially during summer months. Fresh water availability and its management should be the top priority for India.

Case Study 1: A press notification of Government of Odisha shows the action taken in Odisha in connection with WASH.

The Baseline Survey carried out in 2012-2013 and the data shows that only 2178 villages have been declared defaecation free and from 90, 20, 107 rural house-holds only 10, 38, 127 house-holds have toilet facilities. The recent survey shows that 35.39 percent people in the state use toilets and 64.61% people do not have toilet facility. Odisha has 51, 502 villages and 6210 village Panchayat.

To improve the situation, government of Odisha has decided to make Odisha free of open defaecation by 2019 and have taken up many programmes including providing toilet facility in rural hose-holds and urban slums including Community Led Total Sanitation (CLTS). The problem is complex and there is no adequate toilet facility near bus stands, in any place in the highways, market areas, schools and colleges, especially for women. Even if these facilities are provided in future, their maintenance requires priority. The mind-set of people and robust sensitization programmes involving NGOs, School children and general public are to be taken up. Prime Minister's Call for Swachha Bharat is very important and requires a mass movement, adequate finance and committed involvement of the bureaucracy. Let us hope things will improve and we will achieve the desired goal before 2030AD.

References

Wikipedia, the free encyclopedia

UN Water. 2016. 'WWD 2016 facts and guidance'

Bangladesh Bureau of Statistics (BBS) and UNICEF. 2015. Bangladesh Multiple Indicator Cluster Survey (MICS) 2012-2013.

Hunter PR, Prüss-Ustün A (2016). Have We Substantially Underestimated the Impact of Improved Sanitation Coverage on Child Health? A Generalized Additive Model Panel Analysis of Global Data on Child Mortality and Malnutrition. PLoS ONE 11 (10) : e0164571. doi:10.1371/journal.pone.0164571.

http://dx.doi.org/10.1371/journal.pone.0164571

Press Notification of Government of Odisha on 17.11/2016.

World Health Organization and UNICEF Joint Monitoring Programme (JMP). 2015 Progress on Drinking Water and Sanitation, 2015 Update and MDG Assessment.

World Health Organization-UNICEF. 2012. 'Global costs and benefits of drinking-water supply and sanitation interventions to reach the MDG target and universal coverage.'

World Water Assessment Programme, UNESCO. 2015. Water for Women: Every woman counts. Every second counts.

Index

C

H

L

O

T

Y

Z